AQA
GCSE

AIM HIGH
FOR THE TOP GRADES

Biology

£1.99
2
21/4

Muriel Claybrook • Keith Hirst • Sue Kearsey

with Rob Wensley

D0294116

www.pearsonschoolsandfe.co.uk

✓ Free online support
✓ Useful weblinks
✓ 24 hour online ordering

0845 630 33 33

Series editor
Nigel English

Longman
Part of Pearson

Longman is an imprint of Pearson Education Limited, Edinburgh Gate, Harlow, Essex, CM20 2JE.

www.pearsonschoolsandfecolleges.co.uk

Text © Pearson Education Limited 2011
Edited by Stephen Nicholls
Typeset by Tech-Set Ltd, Gateshead
Original illustrations © Pearson Education Ltd 2011
Illustrated by Tech-Set Ltd, Geoff Ward, Tek-Art
Cover design by Wooden Ark
Cover photo: A succulent adapted to a dry environment, with spiny leaves to deter animals seeking food. © Corbis: Micha Pawlitski.

The rights of Muriel Claybrook, Keith Hirst and Sue Kearsey to be identified as authors of this work have been asserted by them in accordance with the Copyright, Designs and Patents Act 1988.

First published 2011

15
10 9 8 7 6

British Library Cataloguing in Publication Data
A catalogue record for this book is available from the British Library

ISBN 978 1 408253 74 8

Advisory Board

Acknowledgements

The authors and publisher would like to thank the following individuals and organisations for permission to reproduce photographs:

(Key: b - bottom; c - centre; l - left; r - right; t - top)

viii PhotoDisc: Life File / Andrew Ward. **1** Pearson Education Ltd: Martin Sookias. **2–3** Science Photo Library Ltd: Nancy Kedersha / UCLA. **5** Shutterstock.com: Ingrid W. (b); Herbert Kratky (t). **8** iStockphoto: gladiolus. **9** Science Photo Library Ltd: Kent Wood (b); Hybrid Medical Animation (t); Dr Linda Stannard, UCT (c). **10** Press Association Images: AP. **12** Corbis: Image 100 (l). Science Photo Library Ltd: CNRI (r). **14** Press Association Images: Kirsty Wigglesworth / PA Wire (b). Rex Features: (t). **18** Science Photo Library Ltd: Steve Percival (b). Shutterstock.com: Brasiliao (t). **20** Shutterstock.com: Stephen Mcsweeny (r); Stephen Mcsweeny (l). **22** Shutterstock.com: Lowe Llaguno. **28** Alamy Images: Nigel Cattlin (l). DK Images: (b). **30** Alamy Images: Country Calm / Den Reader (t). Science Photo Library Ltd: Nigel Cattlin (b). **31** DK Images: Dave King. **35** Science Photo Library Ltd: Larry Dunstan. **36** Press Association Images: Steve Parsons / PA Wire. **37** Photolibrary.com. **38** Shutterstock.com: ostill. **40** Alamy Images: Interfoto. **41** Shutterstock.com: Jiri Miklo (b); dundanim (t). **48–49** Alamy Images: ArteSub. **50** Brand X Pictures: Photo 24 (l). Susan Kearsey: (r); (c). **51** Shutterstock.com: LouLouPhotos (r). Susan Kearsey: (l). **52** Corbis: (b). Creatas: (t). **53** Creatas: (tr). Digital Stock: (br). Getty Images: David Tipling (l). **54** Digital Vision: (tl). Pearson Education Ltd: Tudor Photography (bl). Susan Kearsey: (r). **55** Digital Vision. **56** Science Photo Library Ltd: Dr Ken Macdonald (b). Shutterstock.com: Qing Ding (t). **57** Pearson Education Ltd: Richard Smith (r). Science Photo Library Ltd: Thierry Berrod, Mona Lisa Production (l). **59** PhotoDisc: Photolink / J. Link (b). Shutterstock.com: Tom Curtis (t); Ainars Aunins (r). **60** Shutterstock.com: svic (r). Susan Kearsey: (l). **61** iStockphoto: Nancy Nehring (l). Science Photo Library Ltd: Martin Shields (r). Shutterstock.com: Knorre (c). **64** Shutterstock.com: Matthijs Wetterauw. **65** Corbis: Vienna Report Agency / Sygma (l). Shutterstock.com: ArnoldW (r). **66** Alamy Images: Realimage (r). Science Photo Library Ltd: Robert Brook (l). **67** Shutterstock.com: Jeff Gynane. **69** iStockphoto: Gordon Dixon. **70** PhotoDisc. **72** Warren Photographic. **73** Photos.com. **74** Creatas: (t). Photoshot Holdings Limited: NHPA (b). **75** Science Photo Library Ltd: Brian Bowes. **76** Science Photo Library Ltd: Sinclair Stammers. **78** Science Photo Library Ltd: Eye of Science. **79** Photolibrary.com: BSIP Medical (t). Shutterstock.com: Jaroslav74 (b). **80** Alamy Images: Nigel Cattlin. **81** Rex Features: Philippe Hays (b). Science Photo Library Ltd: Bill Barksdale / AGSTOCKUSA (t). **84** PhotoDisc: (r); (l). **85** PhotoDisc: C Squared Studios. **86** Alamy Images: Interfoto. **90** Corbis: Anthony Bannister; Gallo Images. **91** Digital Vision: (b). Shutterstock.com: Gail Johnson (t). **98–99** iStockphoto: siun. **102** Photolibrary.com. **103** Science Photo Library Ltd: Power and Syred. **104** Shutterstock.com: sgame. **105** Shutterstock.com: Neil Webster (t); KariDesign (b). **106** Pearson Education Ltd: Trevor Clifford (l, c, r). **108** Science Photo Library Ltd: Dr Yorgos Nikas. **110** Science Photo Library Ltd: Cordelia Molloy (br). Shutterstock.com: Marie C. Fields (tr); knin (l). **112** Shutterstock.com: martellostudio (t); Dmitriy Kuzmichev (b). **114** Science Photo Library Ltd: Dr Jeremy Burgess. **115** Shutterstock.com: SamTan (b); nito (t). **116** Shutterstock.com: cybervelvet (r); corepics (l). **117** Shutterstock.com: Brett Nattrass. **118** Alamy Images: Nigel Cattlin. **119** Alamy Images: Bernhard Classen. **124** Shutterstock.com: Thomas Payne. **125** PhotoDisc:

David Buffington. **126** Corbis: Willard Culver / National Geographic Society. **134–135** Science Photo Library Ltd: Herve Conge, ISM. **136** Pearson Education Ltd: Jules Selmes. **140** Pearson Education Ltd: Jules Selmes (t). Shutterstock.com: Jubal Harshaw (b). **142** Alamy Images: Dynamic Graphics. **144** Imagestate Media: John Foxx Collection (r). Shutterstock.com: Carmen Steiner (l). **146** Digital Vision. Guillaume Dargaud: (br). Pearson Education Ltd: Tudor Photography (tl); Malcolm Harris (tr). **147** Shutterstock.com: MichaelTaylor. **149** Science Photo Library Ltd: BSIP, Laurent / B. Hop Ame. **150** iStockphoto: Florian Batschi. **151** Shutterstock.com: Ivan Histand. **154** Science Photo Library Ltd: CNRI. **158** Alamy Images: Momentum Creative Group. **162–163** Corbis: David Woo. **164** Corbis: Jorge Z. Pascual / epa (b). Science Photo Library Ltd: Simon Fraser / RVI, Newcastle-upon-Tyne (t). **166** Photos.com: Jupiterimages (bc). Science Photo Library Ltd: SILKEBORG MUSEUM, DENMARK / MUNOZ-YAGUE (tc). Shutterstock.com: Snowshill (t); mikeledray (b). **167** image courtesy of NASA Earth Observatory: Ames Research Center. **168** Digital Vision. PhotoDisc: InterNetwork Media Inc. (b). **170** Shutterstock.com: jordache (t); Andreas Gradin (b). **171** Ardea: Pat Morris (b). Shutterstock.com: Vishnevskly Vasily (t). **175** Photos.com: Jupiterimages. **185** Science Photo Library Ltd: D. Phillips (l, r). **187** Imagestate Media: John Foxx Collection. **191** Getty Images: Superstock (t). Photos.com. Shutterstock.com: Leonid Smirnov (b). **192** Shutterstock.com: Jubal Harshaw. **194** Science Photo Library Ltd. **198** PhotoDisc. **203** Shutterstock.com: Andreas JÃ¼rgensmeier. **206** Alamy Images: PCN Photography (b); Jenny Matthews (t). **207** Science Photo Library Ltd: Steve Gschmeissner. **210** Science Photo Library Ltd: Zephyr. **211** Science Photo Library Ltd: Life In View (t); Antonia Reeve (b). **212** Alamy Images: Kevin Wheal (b). Getty Images: Tino Soriano (t). **214** Science Photo Library Ltd: Ted Kinsman. **216** SuperStock. **217** Science Photo Library Ltd: Cordelia Molloy. **218** Alamy Images: Jupiter Images / Brand X (t). Getty Images: AFP (b). **219** Shutterstock.com: Bochkarev Photography. **226–227** Shutterstock.com: Tish1. **229** Shutterstock.com: Rechitan Sorin. **230** Digital Vision. **231** Alamy Images: Green Stock Media (b). Shutterstock.com: Norman Chan (t). **232** Shutterstock.com: EMJAY SMITH. **233** Science Photo Library Ltd: J. G. Paren (t). Shutterstock.com: Niels van Gijn (b). **235** Alamy Images: Jim Parkin. **237** Corbis: Enrique Marcarian / Reuters (b). iStockphoto: alandj (c). Science Photo Library Ltd: Prof. David Hall (t). **240** Digital Vision. **241** iStockphoto: Fertnig. **242** Alamy Images: Phototake Inc. **243** Alamy Images: david pearson. **244** Shutterstock.com: Kletr (l); Igumnova Irina (r); Christopher Elwell (c). **246** Digital Vision. Shutterstock.com: Weldon Schloneger (t). **247** Shutterstock.com: Jean Frooms. **248** Shutterstock.com: Willem Tims (l); tonobalaguerf (r). **249** Shutterstock.com: Iakov Kalinin. **250** Shutterstock.com: 1000 Words. **251** Shutterstock.com: GRISHA. **261** Shutterstock.com: tristan tan.

All other images © Pearson Education

The authors and publisher would like to thank the following individuals and organisations for permission to reproduce copyright material. Every effort has been made to contact copyright holders of material reproduced in this book. Any omissions will be rectified in subsequent printings if notice is given to the publishers.

248 Figure 1: Data with kind permission from the Food and Agriculture Organization of the United Nations. **251** Figure 3: Image courtesy of the World Food Programme.

Introduction

This student book has been written by experienced examiners and teachers who have focused on making learning science interesting and challenging. It has been written to incorporate higher-order thinking skills to motivate high achievers and to give you the level of knowledge and exam practice you will need to help you get the highest grade possible.

This book follows the AQA 2011 GCSE Biology specification, the first examinations for which are in November 2011. It is divided into three units, B1, B2 and B3. Within each unit there are two sections, each with its own section opener page. Each section is divided into chapters, which follow the organisation of the AQA specification.

There are lots of opportunities to test your knowledge and skills throughout the book: there are questions on each double-page spread, ISA-style questions, questions to assess your progress and exam-style questions. There is also plenty of practice in the new style of exam question that requires longer answers.

There are several different types of page to help you learn and understand the skills and knowledge you will need for your exam:

- Section openers with learning objectives and a check of prior learning.
- 'Content' pages with lots of challenging questions, Examiner feedback, Science skills, Route to A*, Science in action and Taking it further boxes.
- 'GradeStudio' pages with examiner commentary to help you understand how to move up the grade scale to achieve an A*.
- 'ISA practice' pages to give you practice with the types of questions you will be asked in your controlled investigative skills assessment.
- Assess yourself question pages to help you check what you have learnt.
- Examination-style questions to provide thorough exam preparation.

This book is supported by other resources produced by Longman:

- an ActiveTeach (electronic copy of the book) with BBC video clips, games, animations, and interactive activities
- an Active Learn online student package for independent study, which takes you through exam practice tutorials focusing on the new exam questions requiring longer answers, difficult science concepts and questions requiring some maths to answer them.

In addition there are Teacher Books, Teacher and Technician Packs and Activity Packs, containing activity sheets, skills sheets and checklists.

The next two pages explain the special features that we have included in this book to help you learn and understand the science and to do the very best in your exams. At the back of the book you will also find an index and a glossary.

Contents

Introduction   iii
Contents   iv
How to use this book   vi
Research, planning and carrying out an investigation   viii
Presenting, analysing and evaluating results   x

B1

How organisms work — 2

1 Healthy bodies
B1 1.1 Diet and exercise   4
B1 1.2 Slimming plans   6
B1 1.3 Pathogens   8
B1 1.4 Defence against disease   10
B1 1.5 Treating and preventing disease   12
B1 1.6 Controlling infection   14
B1 1.7 Vaccination programmes   16
B1 1.8 Keeping things sterile   18

2 Coordination and control
B1 2.1 The nervous system   20
B1 2.2 Controlling our internal environment   22
B1 2.3 Controlling pregnancy   24
B1 2.4 Evaluating the benefits of fertility treatment   26
B1 2.5 Plant responses   28
B1 2.6 Using plant hormones   30
Assess yourself questions   32

3 Drugs: use and abuse
B1 3.1 Developing new drugs   34
B1 3.2 Recreational drugs   36
B1 3.3 Establishing links   38
B1 3.4 Steroids and athletics   40
ISA practice: testing hand-washes   42
Assess yourself questions   44
GradeStudio   46

Environment and evolution — 48

4 Interdependence and adaptation
B1 4.1 Plant adaptations   50
B1 4.2 Animal adaptations   52
B1 4.3 Surviving the presence of others   54
B1 4.4 Extreme microorganisms   56
B1 4.5 The effect of changing environments   58
B1 4.6 Pollution indicators   60

5 Energy
B1 5.1 Energy in biomass   62
B1 5.2 Natural recycling   64
B1 5.3 Recycling issues   66
B1 5.4 The carbon cycle   68
Assess yourself questions   70

6 Variation
B1 6.1 Gene basics   72
B1 6.2 Different types of reproduction   74
B1 6.3 Cloning plants and animals   76
B1 6.4 Modifying the genetic code   78
B1 6.5 Making choices about GM crops   80

7 Evolution
B1 7.1 Evolution of life   82
B1 7.2 Evolution by natural selection   84
B1 7.3 The development of a theory   86
ISA practice: The growth of mould on bread   88
Assess yourself questions   90
GradeStudio   92
Examination-style questions   94

B2

Growing and using our food — 98

1 Cells and cell structure
B2 1.1 Animal building blocks   100
B2 1.2 Plant and alga building blocks   102
B2 1.3 Bacteria and yeast cells   104
B2 1.4 Getting in and out of cells   106
B2 1.5 Specialised organ systems   108

2 Plants: obtaining food and growth
B2 2.1 Photosynthesis   110
B2 2.2 Limiting factors   112
B2 2.3 Uses of glucose produced in photosynthesis   114
B2 2.4 Enhancing photosynthesis in greenhouses   116
        and polytunnels
B2 2.5 Manipulating the environment of crop plants   118
Assess yourself questions   120

3 Communities of organisms and their environment
B2 3.1 Communities of organisms and their environment   122
B2 3.2 Collecting ecological data   124
B2 3.3 Analysing ecological data   126
ISA practice: earthworm distribution   128
Assess yourself questions   130
GradeStudio   132

Understanding how organisms function

134

4 Proteins and their functions

B2 4.1 Protein structure, shapes and functions 136
B2 4.2 Characteristics of enzymes 138
B2 4.3 Digestive enzymes 140
B2 4.4 Enzymes used in industry 142
B2 4.5 Home use of enzymes 144

5 Energy from respiration

B2 5.1 Aerobic respiration 146
B2 5.2 Changes during exercise 148
B2 5.3 Anaerobic respiration 150

Assess yourself questions 152

6 Patterns of inheritance

B2 6.1 Cell division 154
B2 6.2 Differentiated cells 156
B2 6.3 Genes and alleles 158
B2 6.4 Inheriting characteristics 160
B2 6.5 Inheriting disorders 162
B2 6.6 Screening for disorders 164

7 Organisms changing through time

B2 7.1 Fossil evidence . 166
B2 7.2 The causes of extinction 168
B2 7.3 The development of new species 170

ISA practice: carbohydrase enzymes 172
Assess yourself questions 174
GradeStudio 176
Examination-style questions 178

B3

Biological systems

182

1 Exchanges

B3 1.1 Diffusion and osmosis 184
B3 1.2 Sports drinks and active transport 186
B3 1.3 Exchanges in humans 188
B3 1.4 Gaseous exchange in humans 190
B3 1.5 Exchange systems in plants 192

2 Transporting material

B3 2.1 The heart and circulation 194
B3 2.2 Blood vessels 196
B3 2.3 The blood 198
B3 2.4 Stents, artificial heart valves, artificial hearts 200
B3 2.5 Transport systems in plants 202

Assess yourself questions 204

3 Homeostasis

B3 3.1 Staying in balance 206
B3 3.2 The role of the kidneys 208
B3 3.3 Treatment with dialysis 210
B3 3.4 Kidney transplants 212
B3 3.5 Controlling body temperature 214
B3 3.6 Controlling blood glucose 216
B3 3.7 Treating diabetes 218

ISA practice: rates of transpiration 220
Assess yourself questions 222
GradeStudio 224

Humans and the environment

226

4 Humans and the environment

B3 4.1 Human activities produce waste 228
B3 4.2 Tropical deforestation and the destruction 230
of peat bogs
B3 4.3 Environmental effects of global warming 232
B3 4.4 Biofuels containing ethanol 234
B3 4.5 Biogas from small-scale anaerobic fermentation 236
B3 4.6 Generating biogas on a large scale 238
B3 4.7 Collecting biogas from landfill and lagoons 240
B3 4.8 Protein-rich food from fungus 242
B3 4.9 Improving the efficiency of food production 244
B3 4.10 Intensive animal farming 246
B3 4.11 Fish stocks 248
B3 4.12 Feeding the world 250

ISA practice: how does temperature affect fermentation? 252
Assess yourself questions 254
GradeStudio 256
Examination-style questions 258

Glossary 262

Index 273

How to use this book

These two pages illustrate the main types of pages in the student book and the special features in each of them. (Not shown are the end-of-topic Assess yourself question pages and the Examination-style question pages.)

Section opener pages – an introduction to each section

An introductory paragraph to help put what you will be learning into context. There are two section openers for each unit.

A list of the learning objectives you will have achieved by the end of the section.

Test yourself on what you should have learned previously that will help with your understanding of this section.

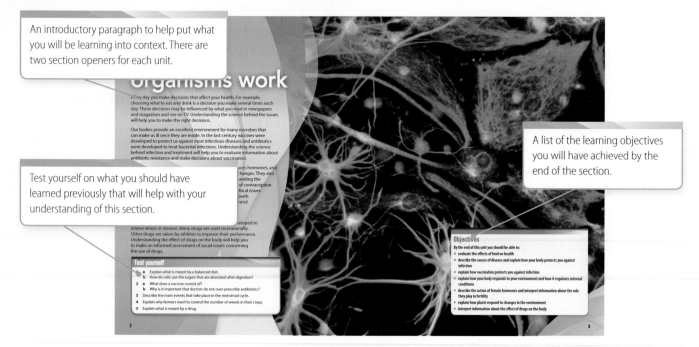

Content pages – covering the AQA specification

A list of objectives for the spread; you can use these to check your progress.

Clear, detailed artwork helps to explain the science.

Keywords are in bold and are listed with their meanings in the glossary at the back of the book to help with revision.

Examiner feedback helps you do better in your exams.

These boxes will help you with your controlled assessment and focus on investigative skills.

Lots of questions at the end of each spread in order of increasing difficulty. The last question on each spread requires a longer answer and is worth six marks.

These boxes highlight specific content or ways to answer questions that will help you get an A* grade.

Science in action boxes highlight new, exciting applications of science.

Taking it further boxes (not shown) cover content that extends from GCSE to A level. You will not be examined on this content but it will provide helpful background.

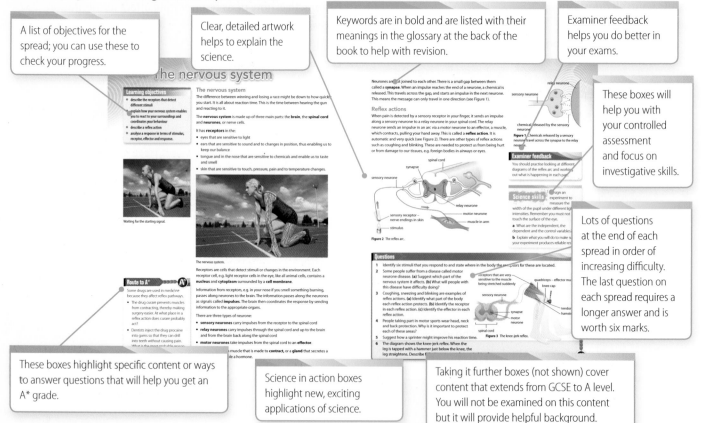

ISA practice pages – to help you with your controlled assessment

The questions are similar to the ones you will be asked in your controlled assessment papers.

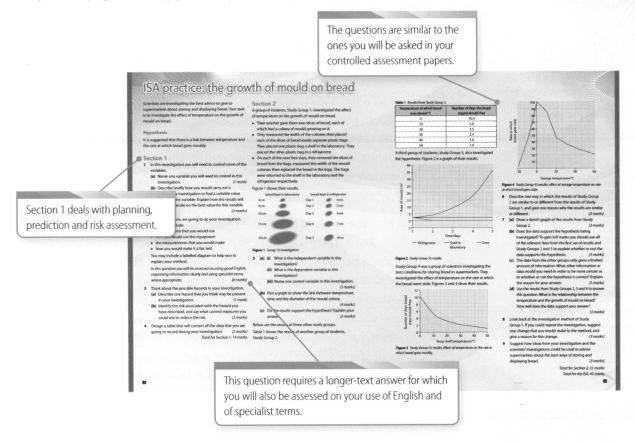

Section 1 deals with planning, prediction and risk assessment.

This question requires a longer-text answer for which you will also be assessed on your use of English and of specialist terms.

GradeStudio pages – helping you achieve an A*

'GradeStudio' questions focus on the new exam questions, which require a longer answer.

Three student answers are given at three different grades, B, A and A*, so you can see how they improve.

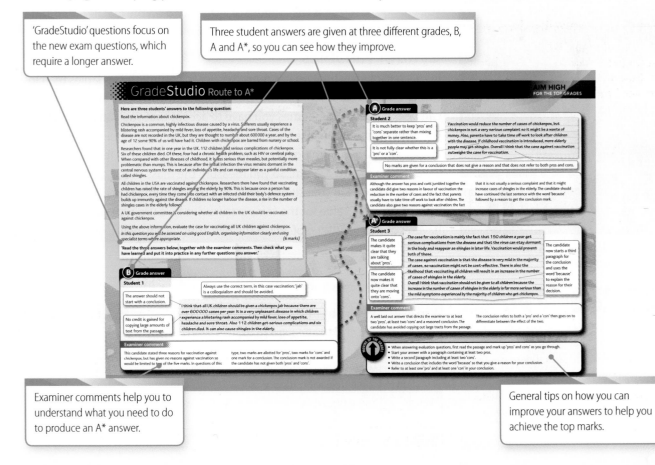

Examiner comments help you to understand what you need to do to produce an A* answer.

General tips on how you can improve your answers to help you achieve the top marks.

Researching, planning and carrying out an investigation

Light levels and leaf sizes vary at different places on a beech tree.

Scientific understanding

Science is all about observing how things behave, trying to understand how they work, and using the understanding in new situations. This understanding must be based on evidence. Ideas and explanations must relate to that evidence, and be able to explain new situations.

Case study: leaves

If you look at a hedge made of beech trees, you may notice that some leaves are bigger than others. You might wonder whether this is just random variation, like the variation in people's heights, or whether there is a reason. You might find that if one side of the hedge faces the sun and the other is in the shade, the leaves on the shady side are larger. Knowing that leaves use sunlight to carry out photosynthesis, you can suggest an explanation or **hypothesis** about the different sizes of the leaves, for example *leaves in shade are larger to allow them to collect the same amount of energy from light as smaller leaves in direct sunlight*. From this hypothesis you can make a prediction to test. One testable prediction would be that the size of the leaf is proportional to the light it receives.

Testing the prediction

To test this prediction, you need to compare the position of the leaf, and thus how much light falls on it, with its size. The **independent variable**, the one you change or select, is the amount of light falling on the leaf. The **dependent variable**, the one that you measure, is the size of the leaf.

Control variables

In this sort of investigation it is much harder to control the variables than inside the laboratory. Table 1 shows some of the **control variables** and how two of them can be controlled. Sometimes you have to accept that you can't keep all variables the same, but you can take them into account when processing the results. As an example, when burning oil to find the energy present in the oil, you cannot always burn exactly 1 gram. So you measure the mass of the oil before and after burning, and divide your temperature rise by the number of grams burnt. This then gives you a value for the temperature rise for burning 1 g, even though you haven't burnt exactly 1 g of any of the oils. Your results now are comparable, and valid.

Table 1 Some variables that are hard to control.

Variable	The problem	How to control it
Light intensity	This changes throughout the day	Measure light intensity every hour, then produce mean
Genetic variation	Leaves from different trees will be genetically different; some trees will produce bigger leaves than others	Sample from only one tree, or use cloned plants/trees
Water	Some trees may have more access to water in the soil, e.g. on river bank	
Nutrients	Different soils contain different nutrients	

Preliminary work

Before you carry out your investigation, you need to be certain the investigation will give you results that are meaningful. Preliminary work allows you to find out the **range** of light intensities that occur and identify control variables.

All scientific investigations have **hazards**, things that can go wrong and cause injury to people or objects. The hazards in this investigation are relatively minor; for example, you might slip in mud while working outside. To minimise the **risk**, or the chance of a hazard happening, **control measures** are used to reduce the hazard to a level of risk that is acceptable. The control measure in this case would be to wear a good pair of boots.

Making measurements

When measuring natural materials like leaves, one way to overcome natural variation is to measure a large number of them. For the leaves you choose, keep as many variables as you can the same, only varying the amount of shade. Choose a method that gives the greatest possible resolution to your measurements. Drawing around your leaf on squared paper, then counting squares, will give poor resolution if the squares are 1 cm², but far greater resolution if the squares are 1 mm². Calculating the mean of the area of the leaves will help to reduce the effect of **random errors**, the small errors in all measurements. These are quite likely if your measurement method involves some degree of estimating, such as counting half-filled squares, but omitting squares less than half filled.

 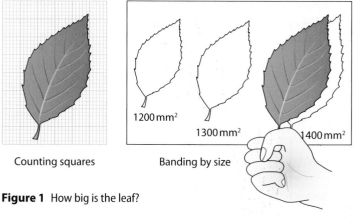

Method A — Counting squares

Method B — Banding by size
1200 mm² / 1300 mm² / 1400 mm²

Figure 1 How big is the leaf?

Counting squares can be time-consuming, so you could choose a less precise method of estimating the size of a leaf that would allow you to size each leaf quickly so you can sample far more leaves. Figure 1 shows two methods of estimating the area of a leaf.

Recording and processing your results

Results charts often have one space for every result, but if you are measuring the area of leaves from a shady site and a sunny site, then a tally chart for each site may be faster – see Figure 2.

Site	Area of leaf/mm²	Tally	Total
Shady	1000	|||	3
	1200					3
	1400	卌 卌	10
	1600	卌 卌		11
	1800	卌 卌	10
	2000	卌 ||||	9

Figure 2 Tally chart for results.

Questions

1. Explain the difference between a hypothesis and a prediction.

2. Suggest how you would control the amount of water in the investigation of beech leaf sizes (Table 2).

3. Suggest how you would control the amount of nutrients in the investigation.

4. Look at Figure 1. If you put leaves into bands according to their area, as in Method B, what average area would you give leaves in the band 1200 mm²–1300 mm²?

5. Suggest how method B in Figure 2 would allow you to size a leaf quickly. Suggest two rules for its use.

6. The amount of light falling on an area can be measured using a light meter. Write a brief plan of how to investigate whether the size of a leaf is proportional to the amount of light falling on it.

Presenting, analysing and evaluating results

graphs showing a linear relationship

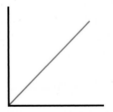

graph showing directly proportional relationship

Figure 1 Examples of line graphs.

Processing your data

On the previous pages you looked at the hypothesis *Leaves in shade are larger to allow them to collect the same amount of energy from light as smaller leaves in direct sunlight.* To judge if the hypothesis is true or false, you need to process the data you have obtained, and look for patterns and trends in the data.

Sometimes you can spot a trend or pattern in a set of results from a table, but more likely you will need to see the data as a bar chart, if the independent variable is **categoric**, or as a line graph, for **continuous variables**.

When plotting a line graph you should first plot the points and then look for a pattern in the points. Draw the best-fit line or curve. Sometimes there are anomalous results, which show up as points that would not lie on your best-fit line or curve. Plot the line or curve leaving these points out.

The best-fit line or curve shows you the relationship between the two variables. Straight lines indicate a linear relationship. In Figure 1, the top graph shows a **positive linear** relationship, while the middle graph shows a **negative linear** relationship. If the line goes through the origin, where the axes meet, then the relationship may be **directly proportional**. This is only the case when the origin of the graph is truly zero on both axes.

Curved lines indicate a more complex relationship. Be aware that if a curved graph has a peak, it can be hard to predict where exactly the peak is for the data you have. You need to get more data from the part of the range where you think the peak is. This would be an improvement to your investigation.

Examiner feedback

In science you should only plot your line with the data you have. Unless you have a value for the origin, do not plot it or connect your line to it. Not all graphs meet the axes at the origin. For example, remember that in winter the temperature drops to less than 0°C. This means that 0°C is not zero for a temperature axis. In fact the lowest possible temperature, the true value for zero, is −273.15°C.

Examiner feedback

When asked for improvements to an investigation you should clearly identify what you want to change, with a clear reason for the change and how this will help to prove or disprove the hypothesis. For example, 'From my results I cannot be sure where the peak activity for an enzyme is in terms of the pH, so I need to obtain more values between pH 4 and 6.'

Analysing the evidence

Your conclusion must relate to the investigation. In our example of leaf area, you should clearly relate the light intensity to the area of the leaf, for example '*My graph shows that as the light intensity increases, the average leaf area decreases.*' You should also state whether this confirms or rejects the hypothesis.

Other people have also researched this type of problem. Football clubs find it difficult getting grass to grow on the pitch because of the shadows cast by the stands. Data from a study into grass-blade area in a football ground could provide data to back up your hypothesis. Look for ways that the data can confirm (or contradict) the trend you have observed. Remember the pattern is the important fact, not the actual figures.

Table 1 and Figure 2 contain more data about leaf area and growth rates. Try to identify which data would support the hypothesis and why.

How would you help this grass to grow evenly?

Questions

1 Look at the graph in Figure 2. Describe as fully as you can the relationship between the light intensity and the length of stalk.

2 Look at the data in Table 1. What evidence is there from Trial 1 to support the hypothesis?

Table 1 The area of hornbeam leaves against light intensity measured for different parts of a hedge.

Light intensity reading/mV	Area of each leaf/cm²			
	Trial 1	Trial 2	Trial 3	Mean area
195	26	28	22	25
408	19	19	22	20
598	14	27	16	19
812	12	8	10	10

3 Look at the data in Table 1 and Figure 2. Does the evidence here support the hypothesis?

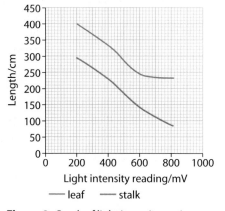
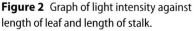

Figure 2 Graph of light intensity against length of leaf and length of stalk.

4 Critically evaluate the evidence from Tables 1 and 2 and Figure 2 and say whether you believe the hypothesis is proved.

Table 2 Light intensity on the pitch at Wembley at 6.00 pm.

Light intensity reading/mV	Position of grass in stadium
857	in front of North Stand
150	in front of West Stand
895	in front of East Stand
524	in front of South Stand

5 If you were to carry out this investigation, suggest improvements to the investigation in terms of measurements and techniques. You should also state any further work you might want to carry out to extend the investigation.

6 The head groundsman at Wembley would like your advice about how to make the blades of grass the same area across all the pitch. On the basis of the data and the investigation, what advice would you give him? Explain why you are giving that advice.

How organisms work

Every day you make decisions that affect your health. For example, choosing what to eat and drink is a decision you make several times each day. These decisions may be influenced by what you read in newspapers and magazines and see on TV. Understanding the science behind the issues will help you to make the right decisions.

Our bodies provide an excellent environment for many microbes that can make us ill once they are inside. In the last century vaccines were developed to protect us against most infectious diseases and antibiotics were developed to treat bacterial infections. Understanding the science behind infection and treatment will help you to evaluate information about antibiotic resistance and make decisions about vaccination.

Two body systems – the endocrine system, which produces hormones, and the nervous system – enable us to respond to external changes. They also help us to control conditions inside our bodies. Understanding the science underlying the use of hormones in some forms of contraception and in fertility treatments will enable you to consider ethical issues concerned with these treatments. Hormones control growth in plants. Understanding the science underlying this control will help you to assess the use of plant hormones in agriculture and horticulture.

Drugs affect our body chemistry. Medical drugs are developed to relieve illness or disease. Many drugs are used recreationally. Other drugs are taken by athletes to improve their performance. Understanding the effect of drugs on the body will help you to make an informed assessment of social issues concerning the use of drugs.

Test yourself

1. **a** Explain what is meant by a balanced diet.
    b How do cells use the sugars that are absorbed after digestion?
2. **a** What does a vaccine consist of?
    b Why is it important that doctors do not over-prescribe antibiotics?
3. Describe the main events that take place in the menstrual cycle.
4. Explain why farmers need to control the number of weeds in their crops.
5. Explain what is meant by a drug.

Objectives

By the end of this unit you should be able to:

- evaluate the effects of food on health
- describe the causes of disease and explain how your body protects you against infection
- explain how vaccination protects you against infection
- explain how your body responds to your environment and how it regulates internal conditions
- describe the action of female hormones and interpret information about the role they play in fertility
- explain how plants respond to changes in the environment
- interpret information about the effect of drugs on the body.

Diet and exercise

Taking it further

Insulin controls the amount of glucose in the blood. In one form of **diabetes**, type 2 diabetes, the body still makes insulin, but needs more than it produces, or cannot use insulin properly. Type 2 diabetes is much more common than type 1 diabetes (in which no insulin is produced). It usually develops in people over the age of 40. The risk of developing diabetes increases with body mass; it is three times more common in people who are 10 kg overweight. However, type 2 diabetes is now being reported in children, all of whom so far have been overweight.

Figure 2 These are the proportions of different foods that make up a balanced diet. Does this match the food that you eat?

Changing lifestyles

The newspaper article in Figure 1 shows how medical experts are worried about the effects of poor diet and lack of exercise on young people. Experts are warning that many children are overweight or obese and therefore have more chance of developing serious health problems in later life.

Science skills

BRITISH CHILDREN TOP LEAGUE FOR UNHEALTHY LIVING

From our health correspondent.

The largest study of youth health reveals that British children live on sugary snacks and almost no fruit and vegetables.

The World Health Organisation report, based on surveys of more than 160 000 children in 35 countries, found that the dietary habits of Britain's young were among the worst.

Doctors recommend at least five portions of vegetables daily for a healthy diet. More than two-thirds of children aged 11–15 admitted that they did not eat even a single portion of vegetables a day. A third of 11-year-olds drank at least one sugary drink a day, as well as eating sweets and chocolate every day. Snacks and sugary drinks are high-energy foods.

Poor diet and increasingly inactive lifestyles are blamed for a massive increase in the number of people who are overweight.

Figure 1
Unhealthy children.

a Use information from the newspaper article to explain why poor diet and inactive lifestyles are blamed for the increase in obesity.

You are what you eat

The amount and type of food that you eat has a major effect on your health. **Carbohydrates** and **fats** in your food provide the energy you need to stay alive and be active. Food also provides the **proteins**, **vitamins** and **minerals** your body uses to grow and to replace damaged cells and tissues. By eating a varied diet you are more likely to get everything you need to keep your body healthy.

Malnutrition happens when you eat the wrong amount of each type of **nutrient**; either too much or too little. **Deficiency diseases** can occur when the body doesn't get enough of a certain vitamin or mineral. These diseases are avoided by eating the right kinds of food. Figure 2 shows the types and proportions of different foods that make up a healthy **balanced diet**.

Metabolism

The more exercise you do, the more energy you need. Many chemical reactions take place in your body. These reactions release the energy from food. Your **metabolic rate** is the speed at which your body uses that energy.

Even when you are resting, you need energy to keep your heart beating and for breathing and digestion. The rate at which you use energy when you are resting is called your resting or basal metabolic rate.

The higher your metabolic rate, the more energy you use. As you exercise your body builds up muscle tissue. Regular exercise also increases your metabolic rate.

Your body uses some of its fat stores to replace the energy that it uses. Exercise increases the proportion of muscle to fat in the body, which in turn increases metabolic rate.

Everybody has a different metabolic rate. It depends on many factors including inheritance and how you live your life. Your metabolic rate decreases with age.

Does your work affect your metabolic rate?

Science skills

b Medical experts recommend that young people carry out at least 60 minutes of moderate exercise every day. What percentage of 15-year-old boys achieve this recommendation?

c Suggest a reason for the change in activity levels among girls as they get older.

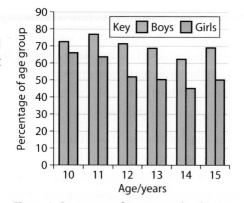

Figure 3 Percentage of young people taking part in at least 60 minutes of physical activity a day.

Blood cholesterol level

Cholesterol is a fatty substance transported by the blood. We all need some cholesterol to keep our body cells functioning normally. Blood cholesterol level depends on the amount of fat in the diet. It is also affected by inheritance. People with high blood cholesterol levels are at increased risk of developing diseases of the heart and blood vessels.

Examiner feedback

Examination questions are likely to provide data that you will be asked to comment upon or evaluate.

Practical

You can compare the energy values of different foods by burning them under a beaker of water as shown in Figure 4. The rise in the temperature of the water is measured for each food. The greater the temperature rise, the greater the energy content of the food.

d Give two variables that should be controlled in this experiment.

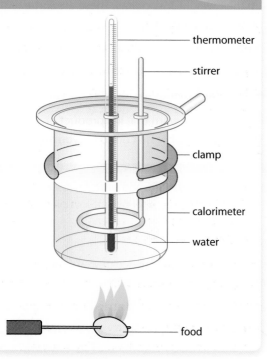

- thermometer
- stirrer
- clamp
- calorimeter
- water
- food

Figure 4 Comparing the energy values of foods using a calorimeter.

Questions

1. Give the function in the diet of: **(a)** fat **(b)** minerals.
2. What is meant by deficiency disease?
3. Give two factors that affect blood cholesterol level.
4. What is meant by metabolic rate?
5. List the factors that affect metabolic rate.
6. Explain the relationship between diet, metabolic rate and body mass.

Slimming plans

New Slimcredible contains 100% natural ingredients that amazingly:

- burn fat
- block hunger pangs from reaching your brain
- boost your concentration
- speed up your metabolism.

Using Slimcredible, you can lose up to one dress size in just two weeks!

Our unique formula contains a blend of guarana, lemongrass, green tea, *Garcinia cambogiam*, L-carnitine and the Acai berry.

Years of scientific studies have shown that this combination of organic ingredients, amino acids and the renowned Brazilian superfood the Acai berry massively boost your weight loss power.

Figure 1 Can you trust all adverts?

Do you believe what you read?

People in the UK spend around £2 billion a year on products that claim to help them lose weight.

Currently, manufacturers are not compelled by law to prove that their products work and can make claims that are unsupported by evidence.

The advert in Figure 1 shows how a fictitious company tries to get people to buy their slimming product as a way of losing mass. The advert tries to convince you that this product works. Many companies use adverts like this. The claims made in the adverts are rarely based on scientific evidence but rather on hearsay. Having an understanding of how evidence is used to provide **reliable** results enables you to question what you read and to make informed decisions.

Medical experts say that there is only one way to lose weight, which is to eat healthy foods and keep energy intake below energy use. Exercise increases the amount of energy used by the body.

Science skills

Adverts showing photographs of a person before and after losing weight by using a slimming product used to be common. This type of advertising has now been banned.

a Suggest reasons why the evidence in this kind of advert should be regarded as unreliable.

Slimming drugs warning

Trading standards officials investigated a variety of slimming products offered for sale over the Internet. They found that three-quarters of the products tested made false claims. Most companies could not provide reliable evidence to back up their weight-loss claims. These are some of the claims that companies make about their products:

- tablets that enable the body to burn fat before food is digested
- pills that allow people to lose weight without dieting or exercising
- a product that burns fat while people are asleep.

Health experts warn that if a product or a diet programme sounds too good to be true, it probably isn't good for you and isn't true.

Science skills

Diet Trials was a study organised by the BBC and a group of scientists.
The study compared four popular commercial weight-loss programmes with a control group.
The diets were:

- the Slim-Fast plan: a meal replacement approach
- Weight Watchers' Pure Points programme: an energy-controlled diet with weekly group meetings
- Dr Atkins' New Diet: a self-monitored low-carbohydrate eating plan
- Rosemary Conley's Eat Yourself Slim diet and fitness plan: a low-fat diet and a weekly group exercise class.

300 overweight people matched for age, sex and mass were randomly divided into the five groups for the study.

Figure 2 shows some of the data from the study.

b Which variables were controlled in the study?

c What feature of the study made it a 'fair test' between the weight-loss programmes?

d Which was the most effective programme in the first two months?

e How did the four programmes compare over the six months?

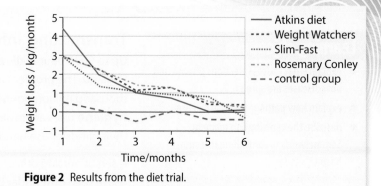

Figure 2 Results from the diet trial.

Science in action

Only one slimming pill, orlistat, has been approved by the NHS.

Orlistat works by inhibiting enzymes that digest fat. Nearly a third of the fat that you eat is not digested if you take orlistat. The undigested fat is passed out with your faeces. Studies have shown that orlistat, plus a mass-reducing diet and exercise, causes more mass loss than a mass-reducing diet and exercise alone. Some people lose 10% or more of their body mass within six months with the help of orlistat. In others, it is less effective.

Questions

1 Explain as fully as you can how a slimming programme such as the Rosemary Conley plan might help a person to lose mass.

2 Look at the four diets in Figure 3.
(a) Compare the ways in which the four diets are intended to make you lose mass. **(b)** Which diet do you think is likely to be most effective? Give reasons. **(c)** Which diet seems to be least credible? Give the reason(s) for your answer.

Atkins Diet

1st stage
• You only eat proteins (meat, fish, poultry, eggs) and fats (oils, butter, etc.).
• You are only allowed 20 g of carbohydrate each day.
• You don't eat any fruit, vegetables or bread. This tricks your body into thinking it is starving and your body uses up its glycogen store.
• Your body loses a lot of water at first.
• If your body is starving it might begin to break down muscle tissue
• It might also damage your kidneys.
Second stage (after a couple of weeks)
• You start to eat more carbohydrate until you stop losing weight.
• This gives you your maximum carbohydrate limit.
• You must stay under that limit to carry on losing weight.

Weight Watchers

• Uses a points-based system.
• Each food is given points based on the amount of fat, fibre and energy it contains.
• Each person is set a points target for the day.
• You can eat anything as long as you do not go over the target.
• You can meet once a week with other 'weight watchers' to measure and discuss progress.

Slim-Fast Diet

• Slim-Fast is a meal-replacement diet.
• You can eat as often as six times a day to avoid highs and lows.
• You take two Slim-Fast shakes for breakfast and for lunch.
• You have a normal dinner.
• Each Slim-Fast meal replacement is around 240 calories and with the meal you should not go over 1200 calories.
• Slim-Fast shakes contain added vitamins and minerals, essential fatty acids and proteins.

Cabbage Soup Diet

• Take a large white cabbage and slice it up. Put it in a pan and cover it with water. Boil it until it becomes a soft pulp. Season to taste.
• You can eat as much cabbage soup as you like.
• It is low in fat but high in fibre.
• You can combine the soup with any fruit or vegetable you like (except for corn, beans, peas and bananas).
• You only stick to this diet for 7 days then move onto a longer term dieting plan.

Figure 3 There are many approaches to dieting.

Pathogens

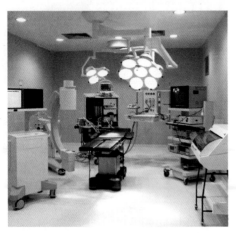

The standard of hygiene in hospitals needs to be high to avoid infections spreading. Think about ways hygiene can be improved in hospitals.

Taking it further

Bacterial cells do not have a nucleus like animal and plant cells. Instead, the genes are found in a looped chromosome and in **plasmids**, small seperate rings of genetic material. Plasmids are used by genetic engineers to transfer genes into both animal and plant cells.

Transmitting infections

Up to 5000 people die each year from infections picked up in hospitals in England. The problem actually affects 100 000 people and costs the NHS a thousand million pounds. It is thought that deadly infections are spread because **hygiene** rules are broken. For example doctors and nurses do not always wash their hands or use hand gel between treating patients.

A senior nursing officer said: 'Levels of cleanliness have deteriorated in recent years. I have seen dust under beds, cotton wool buds on the floor and dirty needles dumped in discarded meal trays. There are guidelines about changing the curtains around beds, cleaning floors and cleaning bathrooms but these are often ignored.'

Potentially fatal infections are carried in dust mites and a study has shown that improving ward cleanliness can reduce infections. Hospitals now employ infection control specialists to reduce the number of infections.

Ignaz Semmelweiss

Microorganisms are the tiny living things that can only be seen using a microscope. They are everywhere, including in the food you eat and inside you. Microorganisms that cause illness or disease are types of **pathogens**.

Ignaz Semmelweiss was a doctor in the mid-1800s. He wondered why so many women died of 'childbed fever' soon after giving birth. He also noticed that student doctors carrying out work on dead bodies did not wash their hands

Science skills

a What was Semmelweiss' observation?

b What was his hypothesis?

The table shows Semmelweiss' original data.

c Why did Semmelweiss calculate the percentage of deaths on each ward rather than relying on the number of deaths?

d What conclusions can be drawn from this data?

Year span	Hospital ward	Number of deaths	Number of patients	Deaths (%)
1833–1838 (same number of doctors and midwives in each ward)	Ward 1	1505	23 509	6.4
	Ward 2	731	13 097	5.81
1839–1847 (medical students and doctors in ward 1; midwives in ward 2)	Ward 1	1989	20 204	9.84
	Ward 2	691	17 791	2.18
1848–1859 (chlorinated hand wash used)	Ward 1	1712	47 938	3.57
	Ward 2	1248	40 770	3.06

Figure 1 Semmelweiss's original data.

before delivering a baby. When he got them to wash their hands in a **chlorinated** hand wash before delivering babies, fewer women died. He concluded that something was carried by the doctors from the dead bodies to the women.

The discovery of pathogens

Louis Pasteur and Joseph Lister studied 'objects' that became known as 'microorganisms'. Pasteur proved that there were '**germs**' in the air and that they carried infection and disease. Lister developed a special soap called carbolic soap. He insisted that all medical instruments, dressings and even surgeons should be cleaned with it before any operation. More of Lister's patients stayed healthy than those of other surgeons.

Chemicals that are used to clean wounds or get rid of sores, such as nappy rash, are called **antiseptics**. Chemicals that are used to clean work surfaces and other places where pathogens might be found are called **disinfectants**.

Semmelweiss showed that keeping things clean helps to stop the spread of pathogens. We call this hygiene. Hygiene is about keeping things clean to reduce the risk of disease. Washing removes the dirt and grease that pathogens stick to and use as a source of energy to multiply.

Microorganisms

Pathogens are the 'germs' identified by Semmelweiss, Pasteur and Lister. Two of the main types are:

- **bacteria** – cause cholera, boils, MRSA, typhoid, tuberculosis
- **viruses** – cause warts, herpes, polio, flu, mumps, measles, smallpox.

Spreading disease

Bacteria and viruses can pass from one person to another. This is how some diseases spread and affect many people. You can become infected by pathogens in the air you breathe, the food you eat, the liquids you drink, and by touching someone. By making sure that your environment is clean, you lessen the chance that you will become infected.

Passing on pathogens.

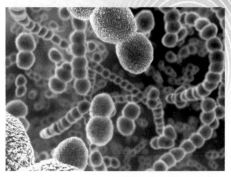

These spherical bacteria are the type that cause sore throats.

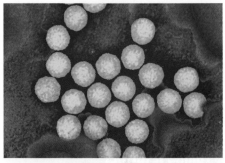

Polioviruses cause the disease polio. They are one thirtieth the diameter of the bacteria above.

Examiner feedback

Do not use the term 'germ' in answers. The general term to use is pathogen. If you use the terms 'bacteria' or 'virus', make sure that they correctly apply to the disease in the question.

Questions

1 What is a pathogen?
2 Name two types of pathogen.
3 Give two ways in which pathogens pass from one person to another.
4 Explain the difference between an antiseptic and a disinfectant.
5 Explain fully why it is important to wash your hands after visiting the toilet.
6 Write a hygiene memo with 10 bullet points for hospital staff. Base your memo on the scientific principles that Semmelweiss applied to his work.

Defence against disease

Immunity

This boy has a condition known as SCID, also called 'bubble boy' disease. He would not survive outside of his 'bubble' because he has no natural defence against pathogens; he is not **immune** to them.

Bacteria and viruses make you ill by releasing poisonous chemicals called **toxins** or by preventing your cells from working properly. You might get **symptoms** such as a headache, fever or feeling sick.

A boy in a bubble.

Once bacteria are inside you, they can multiply rapidly, doubling in number about every 20 minutes. Viruses multiply by entering the cells in your body. They use the chemicals inside the cell to make copies of themselves. The new viruses burst out of the cell ready to invade other body cells. This damages or even destroys the cell.

Cells to fight pathogens

Your body has different ways of protecting itself against pathogens. **White blood cells** are specialised cells that defend your body against pathogens. There are several different types of white blood cell. Some **ingest**, that is, take into the cell, any pathogens that they come across in your body. Once the pathogen is inside the cell, the white blood cell releases **enzymes** to **digest** and destroy it.

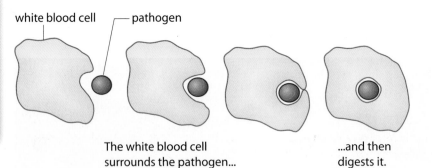

white blood cell — pathogen

The white blood cell surrounds the pathogen... ...and then digests it.

Figure 1 Some white blood cells ingest and destroy pathogens.

Other white blood cells release chemicals called **antibodies**, which destroy pathogens. A particular antibody can only destroy a particular bacterium or virus, so white blood cells learn to make many different types of antibody. For example, when a flu virus enters the body, antibodies are made that destroy the flu virus. After the virus has been destroyed, flu antibodies remain in the blood and act quickly if the same pathogen enters in the future. White blood cells also produce **antitoxins**. These are chemicals that prevent the toxins made by pathogens from poisoning your body.

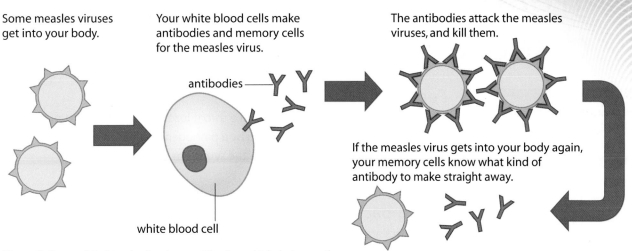

Some measles viruses get into your body.

Your white blood cells make antibodies and memory cells for the measles virus.

antibodies

white blood cell

The antibodies attack the measles viruses, and kill them.

If the measles virus gets into your body again, your memory cells know what kind of antibody to make straight away.

Figure 2 Some white blood cells release antibodies, which destroy pathogens.

Life-long protection

Once your white blood cells have destroyed a type of pathogen, you are unlikely to develop the same disease again. This is because your white blood cells will recognise the pathogen the next time it invades your body and produce the right antibodies very quickly to kill the pathogen before it can affect you. This makes you immune to the disease.

Science skills

Figure 3 shows what happens when someone is infected by a particular pathogen. The graph also shows what happens when the person is infected a second time by the same pathogen.

a How long did it take to start producing antibodies:
  i after the first infection **ii** after the second infection?

b Explain why antibodies were produced more quickly after the second infection.

c Suggest why the person did not become ill after the second infection.

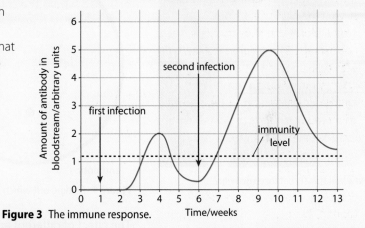

Figure 3 The immune response.

Questions

1 **(a)** Classify the following as B, caused by bacteria, or V, caused by viruses: chicken pox, measles, tuberculosis, mumps, rubella, dysentery, smallpox, cholera, polio, influenza. **(b)** Which of these have you had? **(c)** Which of these have you had only once? **(d)** Which have you had more than once? **(e)** Which are you not immune to?

2 What is meant by immunity?

3 What is meant by a symptom?

4 Explain the difference between antibody and antitoxin.

5 Describe the ways in which pathogens make us feel ill.

6 Describe the ways in which white blood cells protect us against pathogens.

7 Write a paragraph to explain fully why you do not usually suffer from a particular infectious disease more than once.

A*

Treating and preventing disease

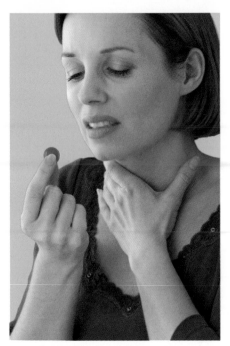

How will this throat lozenge help?

Feeling ill

If you have a sore throat you might take throat lozenges to reduce the pain. The sore throat is a symptom caused by a pathogen that has infected your body. This **medicine** helps to relieve the symptom but it will not kill the pathogen.

Killing bacteria

Antibiotics are medicines that help to cure diseases caused by bacteria. You take antibiotics to kill bacteria that get inside your body. Doctors use many different antibiotics to treat people. **Penicillin** was the first antibiotic to be discovered.

Antibiotics can't kill viruses. Because viruses live and reproduce inside body cells, it is difficult to develop medicines that kill viruses without damaging body cells and tissues.

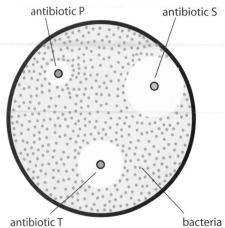

antibiotic P   antibiotic S

antibiotic T   bacteria

Figure 1 Finding out which antibiotic works best.

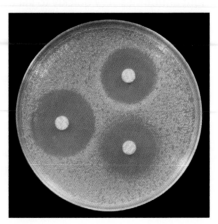

Which antibiotic is most effective?

The effect of different antibiotics on bacteria can be measured in the laboratory. This is done by using small discs of paper containing antibiotics. The discs are placed in a dish containing bacteria growing on a **gel**. The photograph shows the effect of three antibiotics. The clear zone that forms around each disc is where bacteria have been killed. Figure 1 shows the results of testing three other antibiotics. Gel tests are useful, but the body is more complicated than a gel. The results in the body may be different.

A quick jab

Immunity to a disease can be gained without ever having had the disease. A newborn baby receives antibodies from its mother in the first few days that it feeds on her milk. When you were a young child you were probably **immunised** to protect you from very harmful diseases, such as whooping cough, measles and polio. **Immunisation** (**vaccination**) usually involves injecting or swallowing a **vaccine** containing small amounts of a dead or weak form of the pathogen.

Because the pathogen is weak or inactive, the vaccine does not make you ill, but your white blood cells still produce antibodies to destroy the pathogen. This makes you immune to future infection by the pathogen. Your white blood cells will recognise the pathogen if it gets into your body and respond by quickly producing antibodies. The pathogen does not get a chance to reproduce enough to make you ill.

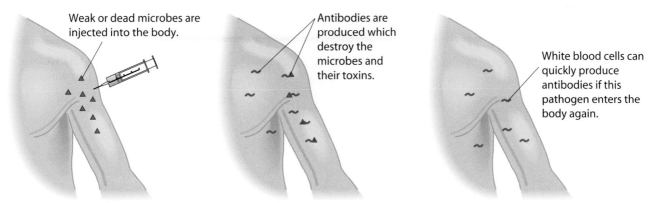

Weak or dead microbes are injected into the body.

Antibodies are produced which destroy the microbes and their toxins.

White blood cells can quickly produce antibodies if this pathogen enters the body again.

Figure 2 How vaccination works.

Science skills

The level of antibody in the blood after some vaccinations does not get high enough to give protection. In this case, a second, or booster, injection of vaccine a few weeks or months later is needed. The graph shows the level of antibodies in a person's blood following a first and second injection of a vaccine.

a What was the difference in arbitrary units in the level of antibody between the first and second injection?

b Explain why the person became immune after the second injection but not the first.

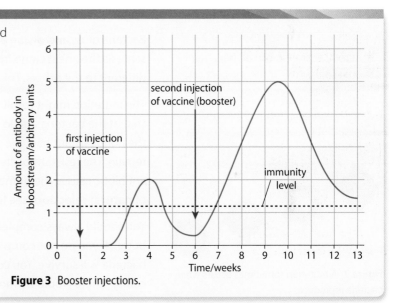

Figure 3 Booster injections.

Questions

1. What symptoms might you have if you catch a cold?
2. What is the most common type of medicine *that is not on these two pages* that is used to treat the symptoms of an infectious disease?
3. Which antibiotic in Figure 1 – P, S or T – works best against this bacterium?
4. Explain why antibiotics can be used to treat bacterial infections but not viral infections.
5. Rubella is a pathogen that can pass across the placenta. Why is it important that girls are vaccinated against rubella?
6. Explain how vaccination protects you against a disease, but does not cause you to develop the disease.

Examiner feedback

It is important that you understand why viruses cannot be killed by antibiotics.

Learning objectives

- explain how strains of bacteria can develop resistance to antibiotics
- explain the consequences of the overuse of antibiotics and of mutations of bacteria and viruses
- describe how epidemics and pandemics occur.

Taking it further

Bacteria may develop antibiotic resistance in several ways. These include mutating to:

- deactivate the antibiotic before it reaches the inside of the bacterial cell
- pump antibiotic out of the bacterial cell
- alter the protein on the bacterial cell so that the antibiotic cannot recognise the cell
- produce enzymes to destroy the antibiotic.

Figure 2 A mutation sometimes occurs when bacteria reproduce.

Dr Semmelweiss's recommendations (lesson B1 1.3) two centuries on.

Antibiotic resistance

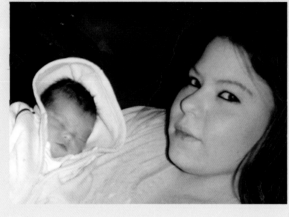

HOSPITAL SUPERBUG KILLS BABY

A one-day-old baby boy was killed by the hospital superbug MRSA.
Baby Luke was only 36 hours old when he died.
Luke was born showing no signs of bad health, but within a day he became ill.

Figure 1 Newspaper article about a hospital 'superbug'.

Bacteria grow and divide every 20 minutes. Each new bacterium is exactly the same as the one it came from. Sometimes a bacterium is produced that is slightly different to the others. This is called a **mutation**.

The mutation might result in the bacteria being **resistant** to existing antibiotics, so that the bacteria are no longer killed by antibiotics. When an antibiotic is used, the non-resistant bacteria are killed but a small number of resistant bacteria remain. The resistant bacteria survive and reproduce. Continued use of the antibiotic causes the number of resistant bacteria to increase. This is an example of **natural selection** (see lesson B1 7.2).

You should always complete a course of antibiotics, even if you start to feel better. If you do not complete a course of antibiotics, it is likely that some bacteria will survive. You could become ill again and need a second treatment course. There would be more chance of resistant bacteria developing.

Scientists are continually developing new antibiotics to replace those that are no longer effective.

Superbugs

Methicillin-resistant *Staphylococcus aureus* (**MRSA**) is a variety of *S. aureus* that is resistant to methicillin and most of the other antibiotics that are usually used to treat bacterial infections. MRSA is responsible for hundreds of deaths of hospital patients every year in the UK.

Patients who die with MRSA are usually patients who were already very ill. It is often their existing illness, rather than MRSA, that is given as the cause of death on the death certificate. In these cases, MRSA is only 'mentioned' on the death certificate.

All UK hospitals now have campaigns to prevent the spread of MRSA.

a Describe the trend in the number of deaths from MRSA.

b Suggest an explanation for the large increase in the number of death certificates where MRSA was 'mentioned'.

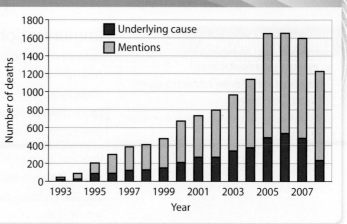

Figure 3 Deaths from MRSA in the UK.

Preventing more superbugs

To prevent more and more types of bacteria becoming resistant, it is important to avoid overusing antibiotics. This is why antibiotics are not used to treat non-serious infections like a sore throat. Doctors should only prescribe an antibiotic to treat a serious disease. By avoiding overusing antibiotics, you increase the likelihood that they will work when you really do need them.

Changing viruses

Flu, or influenza, is a viral disease that affects many people every year. Most people recover within one to two weeks, but flu can cause serious illness and death, especially in very young children and old people. Flu viruses are always mutating, producing new strains. Because the new strain is so different, people will have no immunity to it. This allows the new strain to cause more serious illness and to spread quickly from person to person.

When an outbreak of flu affects thousands of people in a country it is called a flu **epidemic**. Sometimes flu spreads very rapidly around the world, affecting people in many countries. This is called **pandemic** flu. In 2009 a pandemic flu called 'swine flu' developed. The UK government organised vaccination for all vulnerable people and stockpiled millions of doses of Tamiflu, an **antiviral** drug.

Figure 4 An old NHS advert.

Examiner feedback

It is important that you are clear about the difference between an epidemic and a pandemic.

Questions

1 What is MRSA?
2 Name the process that produces antibiotic-resistant strains of pathogens.
3 Explain why patients are advised to always complete a course of antibiotics.
4 Explain why antibiotics are not used to cure viral infections.
5 Suggest why antiviral drugs are only used to treat the most dangerous infections.
6 Explain why the 'swine flu' virus spread rapidly around the world.
7 Explain in terms of natural selection why doctors should not over-prescribe antibiotics.

Science in action

Tamiflu belongs to a family of drugs known as neuraminidase inhibitors. These drugs prevent viruses spreading from cell to cell. When a virus enters a cell it uses the cell's materials to make multiple copies of itself. It also directs the cell to produce an enzyme that will cut through the cell membrane to release the new viruses. Neuraminidase inhibitors block the action of this enzyme. So they do not kill the virus but they do stop it spreading from cell to cell. White blood cells will then eventually develop antibodies to kill the virus.

Vaccination programmes

Immunisation programmes

Immunisation provides protection against several diseases that used to be very common in children. An example is the **MMR** vaccine, a combined vaccine that makes your body develop immunity to measles, mumps and rubella. Each of these diseases is caused by a virus that is easily spread from someone with the disease to someone who is not immune.

Vaccines such as MMR have saved millions of children from illness and even death. Before a measles vaccine was available, an average of approximately 250 000 children developed measles and 85 children died every year, and many others suffered severe symptoms.

If enough people in a community are immunised against certain diseases, then it is more difficult for that disease to get passed between those who aren't immunised. This is because most people who come into contact with the carrier are immunised, so the disease is less prevalent in the population, and therefore even those who are not immunised are less likely to get it. Medical experts recommend that at least 90% of the population should be vaccinated to prevent epidemics of a disease.

Route to A*

About 85% of UK children have been vaccinated against measles. The World Health Organisation has set a vaccination target of 95%. Explain why it is important to reach the World Health Organisation target.

Science skills

a What was the maximum number of cases of measles in any one year before a vaccine against the disease was introduced?

b What was the maximum number of cases of measles in any one year after the introduction of the measles vaccine?

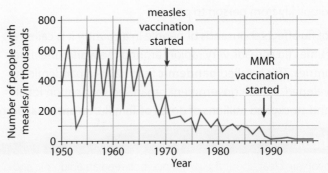

Figure 1 The graph shows the effectiveness of the immunisation programme against measles.

Examiner feedback

Many examinations give data about the number of cases of disease and percentage vaccination.

The data will often have two different y-axis scales. Take time before you answer the questions to familiarise yourself with the key to the two sets of data and the different y-axis scales.

Remember to give arguments both for and against if you are asked to evaluate such data. Conclude your answer with a reasoned conclusion. It is not sufficient to write 'I think the pros outweigh the cons'.

Concern about vaccines

Children who are not vaccinated are much more likely to develop serious illnesses. Whooping cough is a disease that can cause long bouts of coughing and choking, making it hard to breathe. The disease can be very serious and can kill babies under one year old. More than half the babies under one year old with whooping cough need to be admitted to hospital and many need intensive care.

In the 1970s, parents were concerned about possible **side effects** of the whooping cough vaccine and fewer children were vaccinated against whooping cough. As a result, major outbreaks of whooping cough occurred, with thousands of children being taken into hospital.

c Explain why two major outbreaks of whooping cough occurred in 1982 and 1986.

d Describe the relationship between the percentage of children being vaccinated and the number of whooping cough cases since 1990.

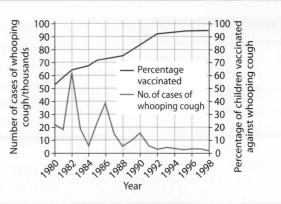

Figure 2 What caused whooping cough epidemics in the 1980s?

The MMR controversy

In 1998, many parents panicked and vaccination rates dropped rapidly after a doctor claimed that the MMR vaccine might trigger **autism**. However, the claim was based on a study of only 12 children. Soon the vaccine was being blamed for the apparent rise in autism in California. In some parts of the UK, the proportion of children receiving the MMR vaccine had dropped to 60% by 2005. This led to a rise in measles outbreaks and fears of an epidemic. Since then, other studies have failed to show any link between autism and MMR. By 2010 the percentage receiving the MMR vaccine had risen to 98%.

Society needs people to be immunised but individual parents can easily be scared off by the risk that their child might react negatively to the vaccine. Balancing risk factors is made more complicated because parents are thinking only of their child, whereas governments are looking at society as a whole.

Examiner feedback

In examination questions asking you to evaluate, make sure that you give your view and that it is supported by the evidence in the article(s).

Questions

1 What is contained in the MMR vaccine?
2 What does MMR stand for?
3 Why is it important to be vaccinated against the three diseases?
4 Why have some parents stopped their children having the MMR vaccine?
5 How would you advise these parents? Explain your answer.

Is there a link between MMR and autism?

The California data in Figure 3 was used by opponents of the MMR vaccine. At first glance it appears to show that an increase in autism is linked to the MMR vaccine. In fact it shows all people registered as having autism in a single year, 1991, plotted by year of birth. The data also do not take into account increases in population in California, nor improved diagnostic measures.

The Yokohama graph shows the number of cases of autism before and after the MMR vaccine was withdrawn.

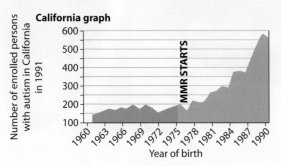

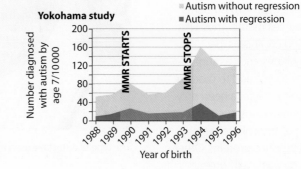

Figure 3 The graphs show data from two studies.

e Do the data in the two graphs indicate a link between the MMR vaccine and autism? Give the reasons for your answer.

Keeping things sterile

Learning objectives

- explain why you need uncontaminated cultures of microorganisms to study the effects of different treatments on them
- describe how apparatus is sterilised and how this achieves pure cultures
- explain how to inoculate a culture medium.

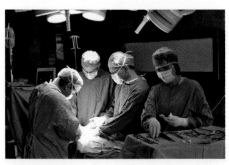

Kitted out to avoid infection.

Examiner feedback

This topic is where your understanding of safety techniques is most likely to be tested. Make sure you can explain the reasons for safety techniques, such as sterilisation, maximum incubation temperature, personal hygiene and disposal of used cultures.

Taking it further

Like animals, most bacteria possess no chlorophyll, so they cannot produce their own food. They cannot eat solid food and have to absorb nutrients through their cell membranes. Like all other living organisms, bacteria use carbohydrates, such as sugars, as an energy source and to produce new cells. Some bacteria need only sugar and mineral ions, such as nitrate, potassium and phosphate, which they use to produce proteins. Other bacteria, particularly pathogenic bacteria, have to obtain proteins and vitamins from external sources.

Growing bacteria

Like all living organisms, bacteria and fungi need nutrients to grow and reproduce. In the laboratory, nutrients are often supplied to the microorganisms, also known as microbes, in a gel called **agar**. Agar is called a growth medium or **culture medium.** It melts at 98 °C and, as a liquid, it can be poured into plastic or glass **Petri dishes**. It solidifies at about 44 °C. Microbes cannot digest agar, so it is not used up as they grow.

Besides nutrients, many microbes need a temperature between 25 °C and 45 °C to grow. In school laboratories, the Petri dishes are put into a cabinet, or an **incubator,** set at a maximum temperature of 25 °C. Pathogens could accidentally be present in the culture dishes, so keeping the temperature at a maximum of 25 °C minimises health risks from them, as they will grow much less well at lower temperatures.

Why do we need to keep things sterile?

The air, the surfaces around you, your skin and clothes all have microorganisms on them. If you culture microorganisms in the laboratory, it involves growing very large numbers of bacterial cells. If safety procedures are not followed, you may accidentally introduce a harmful microbe into a harmless **strain** that you are growing. This would multiply rapidly, just as the harmless microbes do, and would be a greater health risk than if it were a single cell. Sterile or aseptic techniques must therefore be used to prepare uncontaminated cultures.

Glassware and culture media are **sterilised** in an **autoclave** using pressurised steam at a temperature of 121 °C for 15 minutes.

The type of autoclave used in most schools.

The high temperature needed to kill microbes melts plastic, so Petri dishes and disposable instruments are sterilised by **ultraviolet** or **ionising radiation**. This is done commercially. Petri dishes remain sterile inside until the lid is opened.

Safety first

You must wash your hands before and after working with microorganisms. A clean, cleared working surface is also essential. Hair should be tied back, broken skin covered with a plaster and hand-to-face contact avoided while culturing microbes. Work is carried out near the upward draught from a lighted Bunsen burner. The upward movement of air around the burner minimises the risk of airborne microbes falling onto **culture plates**.

Inoculation

Inoculation is the process of transferring microbes to the culture medium. For solid agar, a wire **inoculating loop** is used. It is first sterilised by holding it in a Bunsen burner flame. After cooling the loop for ten seconds, near the Bunsen burner, the microbes can be picked up from the **pure culture** and transferred to the sterile agar by gently sweeping the loop back and forth over the surface.

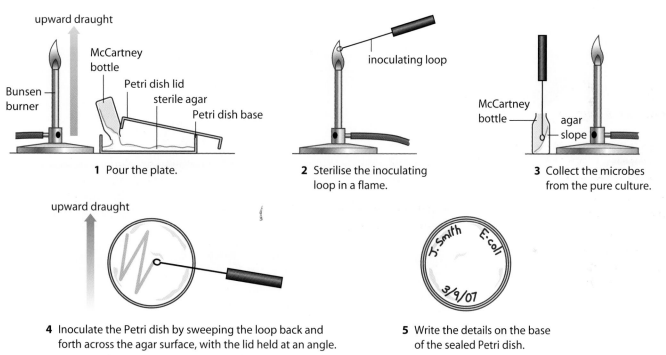

1 Pour the plate.

2 Sterilise the inoculating loop in a flame.

3 Collect the microbes from the pure culture.

4 Inoculate the Petri dish by sweeping the loop back and forth across the agar surface, with the lid held at an angle.

5 Write the details on the base of the sealed Petri dish.

Figure 1 The process of inoculation.

Before **incubation** the Petri dishes are sealed with adhesive tape to prevent contamination from airborne bacteria. After 24–72 hours, when the results have been noted, the cultures are autoclaved by a technician before disposal.

Questions

1 Give three ways in which microbiological equipment and media can be sterilised.

2 What is meant by 'aseptic'?

3 Give three personal safety precautions that should be taken when experimenting with microorganisms.

4 Describe what would happen if a single cell of a pathogen entered a culture dish.

5 Explain why the wire loop needs to be cooled before picking up the microbes.

6 Suggest why the lid of the Petri dish is not removed completely when introducing the microorganisms.

7 Describe what you should do with your successful culture after an experiment.

8 Explain why agar is suitable as a culture medium for growing bacteria.

The nervous system

The nervous system

The difference between winning and losing a race might be down to how quickly you start. It is all about reaction time. This is the time between hearing the gun and reacting to it.

The **nervous system** is made up of three main parts: the **brain**, the **spinal cord** and **neurones**, or nerve cells.

It has **receptors** in the:

- eyes that are sensitive to light
- ears that are sensitive to sound and to changes in position, thus enabling us to keep our balance
- tongue and in the nose that are sensitive to chemicals and enable us to taste and smell
- skin that are sensitive to touch, pressure, pain and to temperature changes.

Waiting for the starting signal.

The nervous system.

Receptors are cells that detect stimuli or changes in the environment. Each receptor cell, e.g. light receptor cells in the eye, like all animal cells, contains a **nucleus** and **cytoplasm** surrounded by a **cell membrane**.

Information from receptors, e.g. in your nose if you smell something burning, passes along neurones to the brain. The information passes along the neurones as signals called **impulses**. The brain then coordinates the response by sending information to the appropriate organs.

There are three types of neurone:

- **sensory neurones** carry impulses from the receptor to the spinal cord
- **relay neurones** carry impulses through the spinal cord and up to the brain and from the brain back along the spinal cord
- **motor neurones** take impulses from the spinal cord to an **effector**.

An effector can be a muscle that is made to **contract,** or a **gland** that secretes a chemical, for example a hormone.

Neurones are not joined to each other. There is a small gap between them called a **synapse**. When an impulse reaches the end of a neurone, a chemical is released. This travels across the gap, and starts an impulse in the next neurone. This means the message can only travel in one direction (see Figure 1).

Reflex actions

When pain is detected by a sensory receptor in your finger, it sends an impulse along a sensory neurone to a relay neurone in your spinal cord. The relay neurone sends an impulse in an arc via a motor neurone to an effector, a muscle, which contracts, pulling your hand away. This is called a **reflex action**. It is automatic and very quick (see Figure 2). There are other types of reflex actions such as coughing and blinking. These are needed to protect us from being hurt or from damage to our tissues, e.g. foreign bodies in airways or eyes.

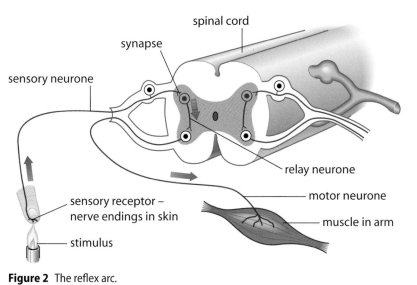

Figure 2 The reflex arc.

Figure 1 Chemicals released by a sensory neurone travel across the synapse to the relay neurone.

Examiner feedback

You should practise looking at different diagrams of the reflex arc and working out what is happening in each part.

Science skills Design an experiment to measure the width of the pupil under different light intensities. Remember you must not touch the surface of the eye.

a What are the independent, the dependent and the control variables?

b Explain what you will do to make sure your experiment produces reliable results.

Questions

1   Identify six stimuli that you respond to and state where in the body the receptors for these are located.

2   Some people suffer from a disease called motor neurone disease. **(a)** Suggest which part of the nervous system it affects. **(b)** What will people with this disease have difficulty doing?

3   Coughing, sneezing and blinking are examples of reflex actions. **(a)** Identify what part of the body each reflex action protects. **(b)** Identify the receptor in each reflex action. **(c)** Identify the effector in each reflex action.

4   People taking part in motor sports wear head, neck and back protection. Why is it important to protect each of these areas?

5   Suggest how a sprinter might improve his reaction time.

6   The diagram shows the knee-jerk reflex. When the leg is tapped with a hammer just below the knee, the leg straightens. Describe fully the sequence of events in this reflex arc.

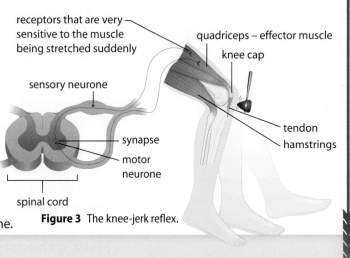

Figure 3 The knee-jerk reflex.

Controlling our internal environment

Learning objectives

- describe how water leaves the body via the lungs, skin and kidneys
- describe how ions are lost through the skin as sweat, and the kidneys as urine
- explain why it is important that blood sugar concentration and blood ion concentration are regulated
- explain why body temperature is kept at 37 °C.

Refuelling during a marathon.

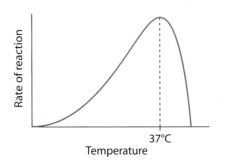

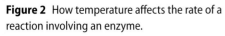

Figure 2 How temperature affects the rate of a reaction involving an enzyme.

When we exercise

During a marathon, runners top up with sports drinks several times. Why do they need to do this?

Figure 1 compares the rate of heat production and the body temperature of a marathon runner during a race with those of the same athlete at rest.

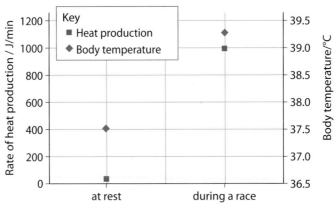

Figure 1 Body changes during a marathon.

Why is our body temperature kept at 37 °C?

Your body tries to stay at a steady internal temperature, around 37 °C. This is the temperature at which enzymes in your body work best. Enzymes speed up chemical reactions in the body. Without enzymes your body would not be able to work properly. The process of keeping things constant and balanced in your body is called **homeostasis**.

Salt (sodium chloride) contains sodium and chloride ions. These are needed to help our body work properly. Too much salt can be dangerous, but so is too little. Sodium and chloride levels in the blood are controlled by the kidneys. These ions are also lost when we sweat.

If the balance of ions and water changes in our bodies, cells do not work so well. Sports drinks help to replace both the water and the ions.

Sports drinks also contain **glucose**. This helps to top up the athlete's blood sugar levels during the marathon. To work properly, body cells need a constant supply of glucose for their energy needs. This glucose is supplied by the blood.

Figure 2 shows the effect of temperature on the rate of an enzyme-controlled reaction. If our bodies were cooler than 37 °C, the chemical reactions in our cells would be much slower. Above 37 °C these reactions rapidly slow down. If we heat enzymes above 45 °C, their structure changes and they stop working.

Heat exhaustion and heatstroke

Heat exhaustion and heatstroke are two heat-related health conditions. Both can be very serious.

Heat exhaustion is when the core temperature rises to 40°C. At that temperature, the levels of water and salt in the body begin to drop. This causes symptoms such as nausea, feeling faint and heavy sweating. If left untreated, heat exhaustion can sometimes lead to heatstroke.

Heatstroke happens when a person's core temperature rises above 40°C. Cells inside the body begin to break down and important parts of the body stop working. Symptoms of heatstroke can include confusion, rapid shallow breathing and loss of consciousness. If left untreated, heatstroke can cause multiple organ failure, brain damage and death.

If a person with heat exhaustion is taken quickly to a cool place and given plenty of water to drink, they should begin to feel better within half an hour and experience no long-term effects. Heatstroke is very serious and should be treated immediately. Treatment involves quickly cooling down the body to lower the core temperature by using ice packs or a cold bath/shower.

Balancing the water budget

To stay healthy, the body needs to balance the gain and loss of both water and ions. Besides losing water when we urinate, pass faeces and sweat, we lose water in the air we breathe out. This is why a mirror becomes misty if we breathe on it.

The **kidneys** control the balance of water and ions in the body. They do this by producing a fluid called **urine**. Urine contains the excess salts and water that the body does not need. It also contains other waste materials.

Science skills

The amount of water entering the body should balance the amount of water leaving the body.

a How much water would this person have to drink to compensate for the amount of water lost?

b What proportion of water loss was via the skin?

c Construct pie charts to show the water budget.

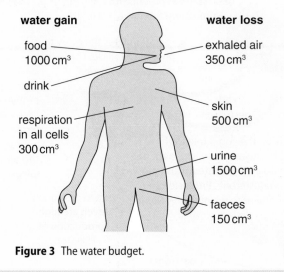

water gain

food 1000 cm³

drink

respiration in all cells 300 cm³

water loss

exhaled air 350 cm³

skin 500 cm³

urine 1500 cm³

faeces 150 cm³

Figure 3 The water budget.

Questions

1 Looking at Figure 1:
   (a) By how much does body temperature rise during a marathon?
   (b) Calculate the percentage increase in heat production by a marathon runner during a race.

2 **(a)** If you sweat a lot, what will happen to your: **(i)** salt levels **(ii)** water levels? **(b)** Why might this be dangerous?

3 Why does an athlete's blood sugar level fall during a race?

4 A person suffering from severe dehydration continues to produce urine, making the body even more dehydrated. Suggest why the body continues to produce urine under these conditions.

5 Many drinks cause the body to lose more water in the urine. These drinks are called diuretics. Alcohol, caffeine and fizzy drinks are all diuretics. Why is it not a good idea to drink these when you feel thirsty?

6 Describe in detail how conditions in the body are kept constant.

Controlling pregnancy

Learning objectives

- explain the role of the hormones FSH, LH and oestrogen in the menstrual cycle
- explain the role of FSH as a fertility drug
- explain the role of hormones in oral contraceptives.

Hormones

Many of our body processes are controlled by chemicals called **hormones.** These are produced by organs called glands. The hormones pass from glands into the bloodstream, which transports them around the body. Each hormone affects one or more organs, known as the target organs.

The menstrual cycle

Every month, an egg (**ovum**) develops inside a female ovary. At the same time, **oestrogen** causes the lining of the womb (**uterus**) to become thicker, ready to receive a growing embryo. If the egg is not fertilised, the womb lining breaks down, causing bleeding from the vagina. The monthly cycle of changes that take place in the ovaries and womb is called the **menstrual cycle**. The menstrual cycle is controlled by several hormones. The action of the hormones involved is summarised in Figures 1 and 2.

Examiner feedback

You are only required to learn the roles of FSH, LH and oestrogen in the menstrual cycle. FSH starts with the letter F. Use this to remember that FSH is the first hormone to be secreted in the menstrual cycle. Remember that FSH stimulates eggs to mature; it does not cause egg release. Oestrogen does not cause egg release. LH starts with the letter L. Use this to remember that LH is the last of the three hormones to be produced. LH stimulates egg release.

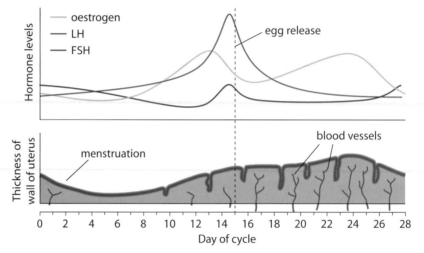

Figure 1 The effects of the menstrual cycle.

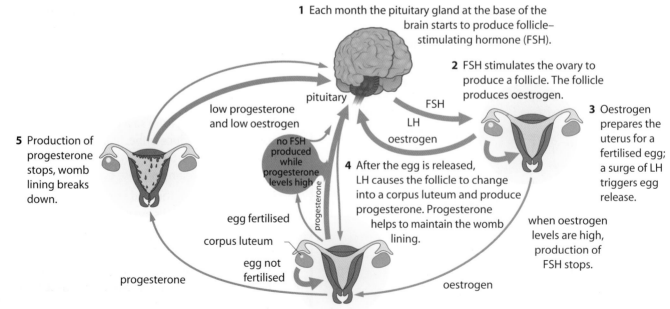

Figure 2 The menstrual cycle.

The contraceptive pill

A woman can take the contraceptive pill to stop her from becoming pregnant. The pill contains hormones that have the same effect on the pituitary gland as oestrogen. These hormones stop the pituitary gland making the hormone **FSH**. This means that no eggs will mature in the ovaries.

Benefits and problems

The first contraceptive pills contained large amounts of oestrogen. These resulted in women suffering significant side effects such as the formation of blood clots, which can block vital arteries.

There are now two types of contraceptive pill available, 'combined' and 'mini pill'. The combined pill contains a much lower dose of oestrogen along with another hormone called progesterone. The mini pill contains progesterone only. The mini pill causes fewer side effects but must be taken punctually in order for it to work and is less reliable than the combined pill.

The combined pill decreases the chance of getting cancer of the womb by 50% and cancer of the ovaries by 40%. There is an increased risk, however, that women taking the pill will develop blood clots.

Fertility drugs

If a woman's own level of FSH is too low, her ovaries will not release eggs and she cannot become pregnant. Infertility can be treated by injecting FSH into the blood. FSH acts as a fertility drug by stimulating the ovaries to produce mature eggs.

Unfortunately, the treatment does not always work, or sometimes it may cause more than one egg to be released. This can result in twins, triplets, quadruplets or even more.

 **Science skills**

a Imagine you are a doctor. Using the data in the table, what would you say to a woman who wanted to go on the pill and was worried about side effects?

Table 1 The risk of blood clots.

Situation	Risk in cases per 100 000 women
women not on the pill	8
women taking combined pill	25
women taking mini pill	15
women who smoke	100
pregnant women	85

Questions

1 **(a)** How many days are there in a typical menstrual cycle?
   (b) On which day does the concentration of FSH peak?

2 What causes menstruation?

3 What is the role of FSH in the menstrual cycle?

4 What is the relationship between oestrogen concentration in the blood and the thickness of the lining of the womb?

5 The most difficult time for a female athlete to race well is the week before menstruation and the week after ovulation. The best time for female athletes to race is thought to be just before ovulation, between days 9 to 12.
   (a) Study the changing levels of oestrogen in Figure 1 and explain the connection between the oestrogen level and athletic performance.

   (b) Design a training schedule for a female athlete so that she can train and race at peak performance. Think about the menstrual cycle and how it affects performance.
   (c) How might taking the contraceptive pill help with race training and performance?

6 Explain why:
   (a) FSH can be used as a fertility drug
   (b) oestrogen can be used as a contraceptive drug
   (c) the mini pill is better in some respects, and worse in others, than the combined pill.

7 There are two forms of 'morning-after' pill (emergency contraceptive pill). Pills containing high doses of oestrogen and progesterone will immediately stop ovulation. The other type of pill prevents a fertilised egg from implanting into the lining of the womb. Evaluate the use of these two types of 'morning-after' pill.

Evaluating the benefits of fertility treatment

Learning objectives

- explain the principles underlying IVF treatments
- evaluate the benefits of, and the problems that may arise from, the use of hormones to control fertility, including IVF.

Interfering with nature?

Sixty-six-year-old Adriana Iliescu became the world's oldest mother when she gave birth to a daughter following fertility treatment.

Ms Iliescu is a retired professor who lives alone. She said she had delayed having a child so she could concentrate on her academic career. Ms Iliescu became pregnant through IVF, using donated sperm and eggs, and this was her third attempt at having a baby.

Figure 1 Interfering with nature?

In vitro fertilisation

Many women are infertile because of blocked oviducts, or fallopian tubes. This means that eggs cannot travel from the ovaries to the womb. Nor can sperm travel upwards to meet an egg. This type of infertility can be treated by using *in vitro* **fertilisation (IVF).** This literally means fertilisation in a test tube, which led to the term '**test-tube baby**'.

The first stage of IVF is to obtain eggs from the woman. She is given injections of FSH to stimulate the **maturation** of several eggs. Eggs are then collected just before they are released from the ovary.

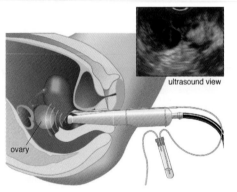
ultrasound view
ovary

Figure 2 Using ultrasound to view the ovary, the doctor inserts the needle through the wall of the vagina into the ovary and removes the eggs for use in IVF.

The eggs are fertilised with sperm from the father outside the body and the fertilised eggs are allowed to divide to form **embryos**, as shown in Figure 3. At the stage when the embryos are still just balls of cells, some are inserted into the woman's womb.

What are the statistics?

The average success rate for IVF treatment using fresh eggs, i.e. eggs that have not been stored frozen, in the UK is shown in Table 1.

The typical cost of a cycle of IVF treatment is approximately £5000–8000. On top of this, the couple will have to pay for the costs of consultation, drugs and tests. The single biggest risk from IVF treatment is multiple births, and particularly triplet births. Many women decide to abort one of these triplets, because multiple births carry potential health risks for both the mother and the unborn children.

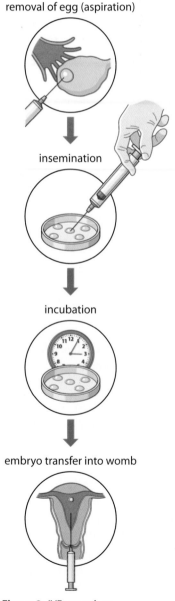

removal of egg (aspiration)

insemination

incubation

embryo transfer into womb

Figure 3 IVF procedure.

Multiple birth babies are more likely to be premature and below normal birth mass. Premature babies have a much higher risk of most of the problems associated with early childhood.

The risk of death before birth or within the first week after birth is more than four times greater for twins and almost seven times greater for triplets than for single births.

The incidence of **cerebral palsy**, a form of brain damage, is approximately five times higher for twins and approximately 18 times higher for triplets than for single births.

Table 1 IVF statistics for 2007 in the UK.

Women aged	Success rate (%)
under 35	32.3
35–37	27.7
38–39	19.2
40–42	11.1
43–44	3.4
44+	3.1

a How is the success of IVF treatment affected by age? Suggest an explanation for this.

b Currently 0.5% of IVF births are triplets, down from almost 4% in the early 1990s. Twenty per cent of IVF births are twins. Suggest an explanation for this.

c Imagine that you are a doctor. What advice would you give to a couple who were considering IVF treatment? Use the information above in your answer.

Science in action

Some couples cannot have children because the man's sperm are too weak to penetrate the egg. These couples may be helped by intracytoplasmic sperm injection (ICSI).

As in IVF, the woman is given fertility drugs to stimulate eggs to mature and to be released. Sperm is collected from the man: a semen sample. A single sperm is then injected directly into the woman's egg with a very delicate needle. The egg will reseal itself after the needle is withdrawn, just as it does in natural fertilisation when the sperm breaks through its outer membrane. The fertilised egg is then allowed to develop for a few days before being transferred back into the woman's uterus in the form of an embryo.

Questions

1 Should Ms Iliescu, Figure 1, have been given fertility treatment? Give arguments for and against.

2 Explain why FSH can be used as a fertility drug.

3 A drug called **clomiphine** is often used instead of FSH. This drug blocks the effect of oestrogen on the pituitary gland. Explain how clomiphine works as a fertility drug.

4 Two embryos are sometimes inserted into the mother, even if she only wants one child. Suggest why.

5 Write a leaflet for couples about to have IVF treatment. Describe and explain the procedures involved. Include diagrams where necessary.

6 During normal IVF, a woman undergoes several weeks of hormone injections. The treatment can lead to a condition called ovarian hyperstimulation syndrome, resulting in a build-up of fluid in the lungs and, very rarely, death. The syndrome occurs in about 1% of standard IVF cycles, but in about 10% for some women. An IVF cycle may cost up to £4300.

In *in vitro* maturation (IVM), hormone treatment lasts for less than 7 days. Eggs are collected from the ovaries while still immature, then matured in a laboratory for up to 48 hours before being injected with a single sperm. A few days after fertilisation, the embryos are implanted into the womb. The cost of each IVM cycle is £1700.

In IVM treatment, the risk of abnormalities in the sex chromosomes, and of birth deformities and cancer in the babies, while small, is greater than in IVF.

Evaluate the use of IVM rather than IVF in treating infertility.

Science skills

Think about the ethical implications of IVF. Here are some opinions. Some people are concerned that it is not natural; others think that an older mother will be unable to cope when the child is a teenager. Some people are concerned about the rights of a woman to be a mother against those of unused embryos. Other people ask whether enough is known about the long-term effects of these techniques. Many people ask why it is acceptable for older men to become fathers, when for an older woman to become a mother is seen as unacceptable.

Plant responses

Learning objectives

- describe the responses of plant roots and stems to gravity, light and water
- explain these responses to light and gravity in terms of the distribution of hormones.

Plant stems grow towards the light.

Phototropism

Plants need light for **photosynthesis**, but they are rooted in soil so they cannot move from place to place to obtain maximum light.

The photograph shows seedlings that have been grown in three different conditions:

- grown in light coming from the left
- grown in darkness
- grown in all-round light.

The stems in the right-hand pot have grown normally.

The stems in the centre pot have grown straight up, and are much longer than the stems in the other two pots because they received no light.

The stems in the left-hand pot have grown towards the light. This response to directional light is called **phototropism**. Plants stems are **positively phototropic**, that is they grow towards the light stimulus.

Which part of the stem detects the stimulus?

Charles Darwin did some of the earliest experiments on phototropism. Figure 1 shows one of them.

This experiment showed that the tip of the stem is the receptor for the light stimulus.

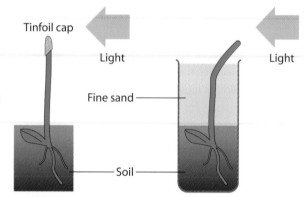

Figure 1 Darwin's experiment.

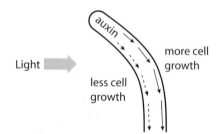

Figure 2 The effector in phototropism.

What is the effector in phototropism?

The response of growing towards the light is brought about by unequal growth. Cells on the shaded side grow longer than cells on the side nearest the light. Growth of cells in the stem is stimulated by hormones called **auxins**. Auxins are produced by the stem tip and are transported downwards. If the stem is placed in **unidirectional** light, more auxin is transported down the shaded side so the cells on this side grow faster, thus bending the stem towards the light.

Gravitropism

The pot shown in the photograph on the left was placed in the vertical position until the stem was a few centimetres tall. The pot was then placed on its side. One day later the stem had turned to grow vertically as shown.

Stems are **negatively gravitropic** – they grow away from the direction of the force of gravity. Roots are **positively gravitropic** – they grow in the direction of the force of gravity.

Gravitropism.

Growing away from gravity means that a stem beneath the soil will eventually find light. Growing downwards into the soil means that a root will help to keep the plant anchored in the soil.

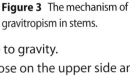

Figure 3 The mechanism of gravitropism in stems.

The mechanism for the gravitropic response of stems is similar to the phototropic response in that it is caused by the unequal distribution of auxin. Auxin accumulates on the underside of a horizontal stem in response to gravity. The cells on the underside grow faster than those on the upper side and the stem grows upwards.

If a root is placed in the horizontal position it will grow downwards. This means that the cells on the upper side grow faster than the ones on the underside. However, auxin accumulates on the underside of the root as it does in the stem. So why does the root go downwards? The reason is that the concentration of auxin that stimulates stem growth inhibits root growth, as shown on the graph in Figure 4.

Using agar and mica

Figure 5 shows one of the experiments that scientists did to show that gravity causes unequal distribution of auxin. Agar is **permeable** to auxins but mica is impermeable.

Hydrotropism

Figure 6 shows what happens when a root is stimulated by both the force of gravity and a directional water stimulus. The root grows towards the water. It is **positively hydrotropic**.

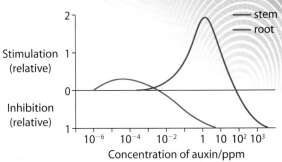

Figure 4 The effect of auxin concentration on root and stem cell growth.

Taking it further

In stems, a photodetector causes auxin-transporter proteins to move into the side membranes of cells. In roots, it is thought that organelles called amyloplasts settle by gravity to the bottom of cells near the root tip. The amyloplasts cause auxin-transporter proteins to move into the cell membranes. These proteins move auxins downwards out of the cells.

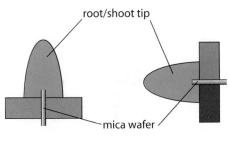

equal amounts of auxin diffuse into both agar blocks

twice as much auxin diffuses into bottom agar block

Figure 5 The distribution of auxins in a gravitropic response.

Questions

1 Name the type of substance that brings about plant responses.
2 Why is mica used in experiments on plant responses?
3 Suggest an explanation, other than the effect of gravity, for the root in Figure 6 growing towards the water.
4 What is the receptor in a tropic response?
5 What is the effector in a tropic response?
6 Name the response of a plant organ to: **(a)** directional light **(b)** the force of gravity **(c)** water.
7 Explain fully why roots grow in the direction of the force of gravity, but stems grow away from the direction of the force of gravity.
8 Compare and contrast the response of a plant shoot to light with a pain withdrawal.

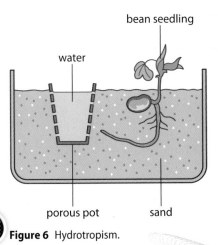

Figure 6 Hydrotropism.

Using plant hormones

The effect of weedkiller.

Selective weedkillers

Most gardens in the UK have a lawn. The biggest problem with lawns is keeping them free of **weeds**. There are two common ways of doing this: pull each weed up by hand or spray the lawn with a **selective weedkiller**.

A selective weedkiller in action on the left.

The most common selective weedkiller is **2,4-D**, short for 2,4-dichlorophenoxyacetic acid.

2,4-D has a chemical structure similar to that of auxins, but it has a much greater effect. 2,4-D is rapidly absorbed by broad-leaved plants. It accumulates in the stem and root tips and causes uncontrolled growth as shown in the photograph.

The growth is so abnormal that the plant dies. 2,4-D is not absorbed by narrow-leaved plants such as grasses, so when it is sprayed on a lawn, the weeds are killed but the grass is unaffected. 2,4-D is a selective weedkiller.

Agent Orange

During the Vietnam war in the 1960s, American aircraft sprayed wide areas of Vietnam with 2,4-D. The codename for this spray was Agent Orange. Agent Orange was an equal mixture of 2,4-D and another hormone, 245T. The concentration of 2,4-D in the spray caused the leaves of the jungle trees to fall off. The purpose of this was to deny hiding places to the North Vietnamese soldiers. Unfortunately the spray came into contact with both Vietnamese and Americans. There was a significant increase in the proportion of children born with birth defects during and after the war. According to the Vietnamese Ministry of Foreign Affairs, 4.8 million Vietnamese people were exposed to Agent Orange, resulting in 400 000 deaths and disabilities and 500 000 children born with birth defects.

Follow-up studies on American soldiers show a higher incidence of several diseases than in the general population.

Rooting powders

Horticulturists use cuttings to produce large numbers of identical plants. Part of a plant shoot is cut off and the end of the stem is dipped in **rooting powder**. The end of the stem is then pressed into damp **compost**. After a few days, roots develop from the cut stem. Rooting powders contain auxins that stimulate the stem cells to develop into roots.

Ripening fruit

The gas **ethene (ethylene)** is a plant hormone. It is produced by fruits as they ripen. One effect of ethylene is to stimulate the reactions that convert starch into sugar. A ripe fruit tastes much sweeter than an unripe one.

Using rooting powder.

Science in action

Bananas grown in the tropics are picked while they are green, which means they are unripe.

They are transported in this state in well-ventilated containers. On arrival at their destination country, the bananas are placed in 'banana rooms'. Ethylene is then pumped into the 'banana room' to ripen them.

Science skills

A student investigated the effectiveness of a rooting powder on three different plant **species**: begonia, geranium and rose. Begonia and geranium are herbaceous (non woody) plants. Rose is a woody bush. The student dipped equal sized pieces of shoots of the three species into rooting powder, then pushed the ends of the stems into damp compost. The student also pushed untreated pieces of shoot from each species into the compost. After two weeks the student measured the total length of roots produced by each species. The results are shown in the table.

a Give two ways in which the student could have improved the investigation.

b Give the two main conclusions that can be drawn from the student's results.

Species	Total length of roots produced/cm	
	Hormone treated shoots	Untreated shoots
Begonia	1.50	0.80
Geranium	0.75	0.40
Rose	0.00	0.00

Practical

Design a controlled investigation to find out if apples ripen faster if stored with bananas.

For 'storage' you could place an unripe apple and a banana into a plastic bag, then seal it.

To compare ripeness of apples, you could cut them in half, then stain with iodine/potassium iodide solution. The more blue/black colour, the more starch.

Questions

1 Name the type of substance used as weedkillers and rooting powders.
2 What is the main difference between the effect of weedkillers on plant stems and that of rooting powders?
3 Suggest why the containers for transporting bananas are well ventilated.
4 Suggest the advantages of transporting bananas in their unripe form, then placing them in 'banana rooms'.
5 Why do selective weedkillers kill only broad-leaved plants?
6 What is the advantage to horticulturists of using rooting powders?
7 Explain the advantages to distributors of using fruit-ripening hormones.
8 How should new weedkillers be trialled before being marketed?

Assess yourself questions

1. Different parts of the body have different functions.
   List A gives three organs.
   List B gives information about each organ.
   Draw one line from each organ in List A to information about the organ in List B.

List A	List B
Organ	Information
gland	produces a fluid that helps to regulate body temperature
kidney	produces hormones
skin	coordinates responses
	produces urine

   (3 marks)

2. A laboratory technician was cleaning out a cupboard. Dust from the cupboard made her sneeze.

   (a) In this response, dust is:
   - A the coordinator
   - B the effector
   - C the receptor
   - D the stimulus

   (b) In this response, the receptor is in:
   - A the brain
   - B the eye
   - C the nose
   - D the spinal cord

   (c) In this response, the coordinator is:
   - A the brain
   - B the nose
   - C the spinal cord
   - D a synapse

   (d) Chemical transmitters are involved in:
   - A sending impulses along sensory neurones
   - B sending impulses across the gap between a sensory neurone and a relay neurone
   - C sending impulses from one end of a relay neurone to the other
   - D sending impulses from a motor neurone to a relay neurone

   (4 marks)

3. Figure 1 shows the reported number of cases in UK hospitals of an infection with a pathogen called *Clostridium difficile*.

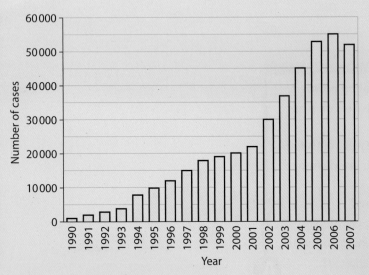

Figure 1 Hospital infections.

(a) Describe as fully as you can the pattern shown by the data in the graph. *(2 marks)*

(b) Suggest an explanation for the change in the number of reported cases of *Clostridium difficile* infection between 1990 and 2005. *(2 marks)*

(c) Suggest an explanation for the change in the number of reported cases of *Clostridium difficile* infection between 2006 and 2007. *(2 marks)*

4. The passage contains information about the 'morning-after' pill.

What does the pill do?
The 'morning-after' pill stops you from becoming pregnant. It is not 100% effective, but the failure rate is quite low – probably about 10%, and rather better than that if you take it as early as possible.

The pill is believed to work principally by preventing your ovaries from releasing an egg, and by affecting the womb lining so that a fertilised egg can't embed itself there.

In Britain and many other western countries, it is not legally regarded as an abortion-causing drug, but as a contraceptive.

Who is the pill for?
It is now very widely used by women (especially young women) who have had unprotected sex. It has proved to be of value to rape victims, couples who have had a condom break and women who have been lured into having sex while under the influence of drink or drugs.

Is it dangerous to use?
You might feel a little bit sick after taking it, but only about 1 woman in 40 actually throws up. Uncommon side effects are headache, stomach ache and breast tenderness.

If the pill didn't work, and I went on and had a baby, could the tablet damage it?
We simply don't know the answer to this question. At present, no one has shown any increase in abnormalities among babies who have been exposed to the morning-after pill. However, past experience does show that other hormones taken in early pregnancy have harmed children.

(a) Some people regard this pill as an abortion-causing drug. Explain why. *(2 marks)*

(b) (i) Some people think that this pill should only be available on prescription. Suggest why they think this. *(1 mark)*

(ii) Others say it should be freely available 'over the counter'. Give two reasons why they think this. *(2 marks)*

(c) Scientists are uncertain if the pill might cause abnormalities among unborn children. Suggest why. *(2 marks)*

5 Figure 2 shows the average amount of cholesterol in the blood of people at different ages.

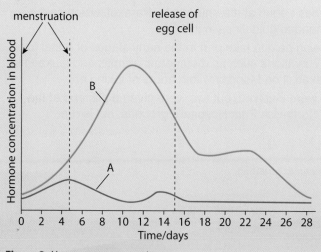

Figure 2 Cholesterol levels.

(a) What is the average blood cholesterol level for a 60-year-old woman? *(1 mark)*

(b) Which group of people has the highest risk of developing heart disease? *(1 mark)*

(c) Give two factors that influence blood cholesterol level. *(2 marks)*

6 Figure 3 shows how the concentrations of the hormones that control the menstrual cycle vary over 28 days.

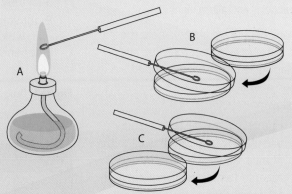

Figure 3 Hormone concentration.

(a) Name:

   (i) hormone A *(1 mark)*

   (ii) hormone B *(1 mark)*

(b) Explain why hormone A can be used as a fertility drug. *(2 marks)*

(c) Hormones similar in their effect to hormone B can be used as contraceptive drugs. Explain why. *(2 marks)*

7 Figure 4 shows the number of cases of influenza in a large city in the UK.

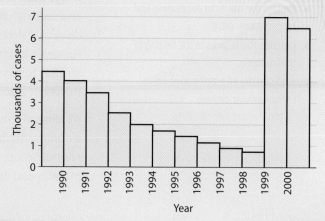

Figure 4 Influenza.

(a) The number of cases of influenza decreased between 1990 and 1998. Suggest an explanation for this. *(2 marks)*

(b) Suggest an explanation for the large increase in the number of cases in 1999. *(2 marks)*

(c) Most people who get flu recover in a few weeks. Explain why. *(2 marks)*

(d) Explain in detail why a person who has been vaccinated against flu may still catch the disease. *(6 marks)*

In this question you will be assessed on using good English, organising information clearly and using specialist terms where appropriate.

8 Figure 5 shows bacteria being transferred from one Petri dish, B, to another, C.

Figure 5 Transfer of bacteria.

Describe fully the procedures you would use with the apparatus shown in the diagram to transfer bacteria from dish B to dish C. Give the reason for each of the procedures you describe. *(6 marks)*

In this question you will be assessed on using good English, organising information clearly and using specialist terms where appropriate.

B1 3.1

Developing new drugs

Learning objectives

- explain the stages in developing and testing a new drug
- describe the failures in the testing of thalidomide
- evaluate the effect of statins on cardiovascular disease.

Developing new drugs

Drugs used to treat disease need to be safe, effective, chemically **stable**, and successfully taken in and removed from the body. The treatment of disease is always being improved by the development of new drugs. However, before a new drug can be used, it is put through several tests and has to pass each stage. As the newspaper extract in Figure 1 shows, this is not always without risk.

All **clinical trials** are **double-blind trials** in which some patients are given a dummy medicine, called a **placebo,** which does not contain the drug, as a control group. Neither the doctors nor the patients know who has received a placebo and who has received the drug being tested, until the trial is complete.

SIX TAKEN ILL AFTER DRUGS TRIAL

Six men remain in intensive care after being taken ill during a clinical drugs trial in north-west London

The healthy volunteers were testing an anti-inflammatory drug at a research unit based at Northwick Park Hospital when they suffered a reaction.

Relatives are with the patients, who suffered multiple organ failure. Two men are said to be critically ill.

An investigation has begun at the unit, run by Parexel, which said it followed recommended guidelines in its trial.

The men were being paid to take part in the early stages of a trial for the drug to treat conditions such as rheumatoid arthritis and leukaemia until they were taken ill on Monday within hours of taking it.

Eight volunteers were involved, but two were given a placebo at the unit, which is on Northwick Park Hospital's grounds, but is run independently.

Figure 1 Drugs tests sometimes go wrong.

Table 1 The main stages in the testing of a new drug.

Stage	Purpose
laboratory	animals or tissues used in a laboratory to find out the level of **toxicity** and to find out if the drug works
phase 1 clinical	low doses are tested on a small group of healthy people to evaluate its safety, and identify side effects
phase 2 clinical	tested on a larger group of people to see if it is effective, to further evaluate its safety and to determine the optimum dose
phase 3 clinical	tested on large groups of people to confirm its effectiveness and monitor side effects

Science in action

Animal testing in the cosmetic industry is especially undertaken for make-up and soaps. Rabbits are the main animals used in these tests. Guinea pigs are used to test sunscreen products. The tests are done to find whether the products will produce reactions and allergies .

Many people are uneasy about the use of animals to test make-up and soaps. Scientists have developed 'non-animal' alternatives including the following.

- Human skin model tests are now in use, including EpiDerm™ (cultured human skin) test, which has been accepted almost universally as a total replacement for skin testing in rabbits
- Human skin left over from operations or from people who have donated their bodies to science is used to measure the rate at which chemicals are able to penetrate the skin.

When drug testing fails

Starting in 1957, **thalidomide** was given to women in the first few months of pregnancy to help them sleep and to overcome the effects of morning sickness. Many women who took the drug gave birth to babies with limbs that weren't properly formed. The drug was banned worldwide in 1961 after it was confirmed that it caused tragic birth defects. The testing of thalidomide was incomplete, because it had not been tested on pregnant animals. The total number of babies damaged by thalidomide throughout the world was about 10 000.

Recently, thalidomide has been used very effectively to treat a serious disease called **leprosy**. However, some pregnant women with leprosy have obtained the drug without doctor's advice, and once again children are being born with deformed limbs.

Testing statins

Drugs called **statins** have been developed to lower blood cholesterol levels.

The result of not testing thalidomide.

 **Science skills**
A study into statins was carried out in a leading UK hospital. The study involved 20 536 patients aged between 40 and 80 with heart disease. The health of these patients was monitored closely over a five-year period.

A total of 10 269 of the patients took a **simvastatin** tablet daily, whilst 10 267 received a placebo every day. Patients were randomly placed into the 'statin' group or the 'placebo' or control group.

The main conclusion of the study was that simvastatin is safe and reduces the risk of people having a heart attack or a stroke.

a This type of study is called a **randomised controlled trial**.

  i What feature of the study was randomised?

  ii How was a control used in the study?

b It is important that studies to assess drugs are highly reliable. What features of this study show that the findings are reliable?

Examiner feedback

Make your own mnemonic to remember the sequence of events in drug testing.

T – tissues for toxicity

H – healthy people for side effects

P – patients for effectiveness

D – patients for dose

Questions

1 **(a)** What conditions was thalidomide designed to treat? **(b)** What test was omitted during the development of thalidomide?

2 What is a double-blind trial?

3 What is a placebo?

4 Suggest why only a few people are used in phase 1 of a clinical trial.

5 Suggest why phase 2 is carried out after phase 1 and not at the same time.

6 Once a drug is on sale, why does it still have to be monitored?

7 Some doctors have used the results of the above statin trial to suggest that all children should be given statins daily. Do you agree with these doctors? Explain the reasons for your answer.

Recreational drugs

What is a drug?

Many drugs are extracted from natural substances. They have been used by people from different cultures for medicines and recreation for thousands of years. **Alcohol** has been fermented from fruit and grain since at least ancient Chinese times, about 9000 years ago.

MAY DAY CELEBRATIONS END IN TRAGEDY

Hundreds of students followed a centuries-old tradition and jumped off Magdalen Bridge at dawn on May Day into the River Cherwell at Oxford.

However, most of them were drunk after all-night parties and they ignored police warnings that the river was too low this year. The result was at least ten students with serious injuries to spines, legs and ankles.

Would students do this if they were sober?

A **drug** is any chemical that alters how our body works. Drugs that affect our **central nervous system (CNS)** control the movement of chemicals across the synapses. The natural chemicals in our nervous system have shapes that fit receptors in our bodies like a key in a lock. Drugs have similar shapes to these chemicals and mimic, or copy, what they do.

People take drugs for recreational or medical reasons. Most drugs were originally used to deal with injury or sickness. Drug abuse occurs when people take too much of a drug or use it for the wrong reasons. Spanish explorers learned from indigenous South Americans to chew on coca plant leaves to keep awake. Today it is used to make cocaine.

If some drugs are used a lot, your body builds up a tolerance to them. This means you must use more of the drug to get the same effect. As a drug is used more often and in greater amounts, your body becomes more dependent on it. This means you will find it difficult to manage without the drug and will need to take it regularly. This leads to addiction. You are addicted when you cannot manage without taking the drug.

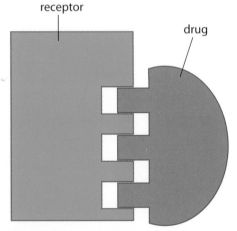

Figure 1 Some drugs mimic chemicals released across the synapses.

When you try to stop taking a drug you are addicted to, you suffer from withdrawal symptoms. These can include feeling sick, headaches and flu-like symptoms. More severe withdrawal symptoms include tremors and fits.

Why do people use drugs?

Drugs are not just taken for medical reasons; some drugs are also taken for pleasure. These drugs are called **recreational drugs**. Some recreational drugs are legal, for example alcohol, **caffeine** and **nicotine** in tobacco, but other recreational drugs are illegal, for example, **cannabis**, **cocaine**, **heroin** and **ecstasy**. Ecstasy, cannabis, cocaine and heroin may have adverse effects on the heart and circulatory system.

Many people smoke cannabis as a recreational drug because it alters their mood. Some of these people think that cannabis is harmless because it is similar to smoking tobacco or drinking alcohol. Many scientists think that the evidence shows that cannabis can cause psychological problems. However, other scientists are uncertain whether cannabis actually causes these problems.

However, it isn't only illegal drugs that can be dangerous. Research has shown that tobacco and alcohol cause thousands of deaths in Britain every year. The NHS has to spend far more money on treating the effects of legal drugs than illegal drugs, because far more people use them. Many people would argue that these drugs should be made illegal. Others argue for the decriminalisation of all drugs, in order to make their use safer, cut down crime associated with the cost of illegal drugs, and to stop criminal gangs from making huge profits from the drugs trade.

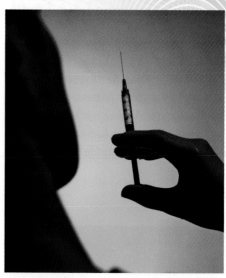

Heroin addicts have to inject themselves frequently or they suffer severe withdrawal symptoms.

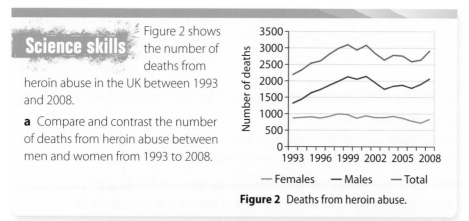

Science skills Figure 2 shows the number of deaths from heroin abuse in the UK between 1993 and 2008.

a Compare and contrast the number of deaths from heroin abuse between men and women from 1993 to 2008.

Figure 2 Deaths from heroin abuse.

Questions

1 Explain what is meant by a recreational drug.

2 **(a)** Give two examples of legal recreational drugs. **(b)** Give two examples of illegal recreational drugs.

3 Explain why drugs can alter the way we behave.

4 'Tonics' sold to people in America in the 1800s contained heroin. **(a)** What would a person feel having taken this 'tonic'? **(b)** What long-term problems might they have suffered?

5 Why might taking heroin lead to you getting HIV, AIDS or other infections?

6 What is meant by 'withdrawal symptom'?

7 Why do more people die from using nicotine and alcohol than heroin and cocaine?

8 Explain fully how a person becomes addicted to a drug.

Establishing links

Many people think that smoking a joint is harmless, but is it?

A harmless joint?

Many people smoke cannabis as a recreational drug; it helps them to 'chill out'. However, there is mounting evidence of a link between smoking cannabis and mental illness. There is also some evidence that smoking cannabis can lead to addiction to hard drugs, in other words, that it is a **'gateway' drug**.

Science skills Figure 1 shows the government classification of drugs and their harm rating compiled by independent experts.

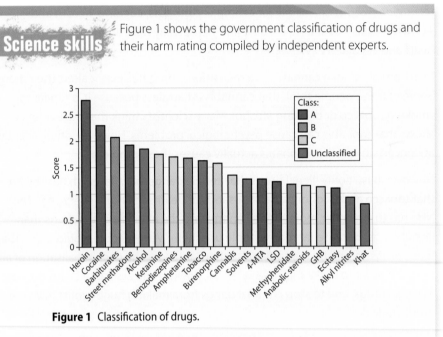

Figure 1 Classification of drugs.

a How does the government classification compare with the harm rating by independent experts?

Is there evidence for a link between cannabis smoking and mental illness?

Patients with mental illness may lose contact with reality or suffer from delusions. Here are some results of research into cannabis and mental illness.

New Zealand scientists followed 1000 people born in 1977 for the next 25 years. They interviewed people about their use of cannabis at the ages of 18, 21 and 25. The questions were about their mental health. The researchers took into account factors such as family history, current mental disorders, and illegal substance abuse.

The scientists' findings were:

- mental illness was more common among cannabis users
- people with mental illness did not have a greater wish to smoke cannabis
- cannabis probably increased the chances of developing mental illness by causing chemical changes to the brain
- there was an increase in the rate of mental illness symptoms after the start of regular use of cannabis.

Scientists studied 45 000 Swedish male conscripts (men called up for army service). This was 97% of the male population aged 18–20 at that time. They followed these men for the next 15 years. They found the men who smoked cannabis heavily at the age of 18 were six times more likely to develop schizophrenia in later life than those who did not smoke cannabis.

Table 1 Drugs studies.

The Amsterdam Study	The Home Office Study	The view of Drugscope
Four surveys, covering nearly 17 000 people, were carried out in Amsterdam in the 1990s. Amsterdam then had 5000 hard drug users in its population of 700 000 and a much larger proportion of cannabis users. There were 300 'coffee shops' in the city where cannabis was freely available.	This study used information from the 1998/99 Youth Lifestyles Survey (YLS), which contains information taken from over 3900 interviews with young people on their own experiences of drug use.	A spokesman for the UK charity Drugscope backed the study's findings. He told BBC News Online: 'Sixty per cent of young people aged 20–24 have used cannabis, but only 1% of that age group have used harder drugs.'
The surveys showed that cannabis users typically start using the drug between the ages of 18 and 20. Cocaine use usually starts between 20 and 25. However, the study concludes that cannabis is not a stepping stone to using cocaine or heroin.	The study found that the age for use of soft drugs is less than the age for use of most hard drugs. However, there was no significant link between soft drug use and the risk of later involvement with crack and heroin. There was a significant but small link between soft drug use and the use of social drugs ecstasy and cocaine.	He said people who used harder drugs were less likely to have 'risk-averse' lifestyles and more likely to have misused other substances, including cannabis, tobacco and alcohol.
The study also claims that most of the evidence that cannabis is a gateway to the use of hard drugs is circumstantial. It found that there was little difference in the probability of an individual taking up cocaine regardless of whether or not they had used cannabis.		

Questions

Regarding the Swedish study:

1 What type of scientific research was this?

2 How can the reliability of surveys be improved?

3 What were the control variables in the investigation?

4 What did the investigation show about the link between cannabis and psychosis?

Regarding the Amsterdam Study, the Home Office Study and the view of Drugscope:

5 Do you think that the data from these studies are reliable? Explain the reasons for your answers.

6 Give two conclusions common to all three studies.

7 Give one possible reason for progression from soft drugs to hard drugs.

8 Consider all the evidence on these two pages. Does using soft drugs lead to taking hard drugs? Explain the reasons for your answer.

Science in action

People may not judge evidence on its scientific strength, but on other criteria. For example, a famous scientist may be taken more seriously than an unknown one, and politicians may disregard scientific advice that they know will be unpopular with voters. We are more likely to accept evidence that agrees with our own views than evidence that contradicts them. Did you dismiss any evidence on this page that disagrees with your opinions?

Steroids and athletics

Learning objectives

- describe how steroids can be used as performance-enhancing drugs in athletics
- evaluate the use of drugs to enhance performance in sport and consider the ethical implications of their use
- describe the effects of using steroids on health.

Science skills

Table 1 shows the percentage of American high-school students who admitted to using steroids to build up muscle in the last year.

Table 1

	Percentage of males	Percentage of females
Social class		
Low	2.2	0.8
Low-middle	1.5	2.3
Middle	2.5	1.2
High-middle	0.6	2.3
High	1.6	0.1
Race		
Native American	9.7	0.0
Other Asian	1.4	0.0
Hispanic	0.0	2.6
Black	0.6	3.7
White	1.3	0.2

a Describe the patterns present in the data for steroid use.

Using steroids to cheat in athletics

Marion Jones winning gold in the 2000 Olympics…

At the 2000 Olympics, sprinter Marion Jones became the first female athlete to claim five medals in a single Games, three of them gold. Eight years later, Jones was headline news again. She was sentenced to serve six months in prison for lying to investigators after admitting her performances in Sydney had been enhanced by **steroids**.

'I have no-one to blame but myself for what I've done,' she said after her admission. 'Making the wrong choices and bad decisions has been disastrous.'

What are steroids?

In ancient times it was known that **testes** were required for the development of male sexual characteristics. In 1849 a scientist called Berthold removed the testes from cockerels, which then lost their sexual function. Then he removed the testes and transplanted them into the birds' abdomens. This time the sexual function of the birds was unaffected. Berthold showed that 'male hormones' passed from the testes into the blood.

In the 1930s, the male hormone was identified as **testosterone**. Scientists were then able to synthesise testosterone and other hormones that acted in the same way. These hormones are called steroids.

In 1954, the Soviet Union dominated the World Weightlifting Championships. They also easily broke several world records. It was soon discovered that the weightlifters had used steroids during their preparation. The use of steroids in athletics had begun. By the 1970s most athletics authorities had banned the use of steroids.

In 1996 scientists investigated the effect of high doses of testosterone on the performance of weightlifters. Forty experienced male weightlifters were randomly assigned to one of four groups: two groups were given a placebo, one group with and one without exercise, and two groups were given testosterone, one group asked to exercise, the other not. The investigation lasted six weeks. The scientists tried to match each group for diet, training and weightlifting experience.

The scientists then measured the strength of each weightlifter and his fat-free body mass, which is the total body mass minus the estimated mass of fat in his body. They found that:

- body mass increased only in the two testosterone-treated groups
- fat-free body mass increased only in the exercise groups
- the greatest change in fat-free mass was in the testosterone plus exercise group
- percentage body fat did not change in any group
- muscle size increased more in the testosterone groups than in either placebo group
- strength increased in both testosterone groups, as well as in the exercise group receiving placebo
- strength increase was greater in the exercise group with testosterone than in the exercise group with placebo.

Other scientific studies have shown that there is a significant **placebo effect** in studies with weightlifters. In most of these studies, strength increased considerably in subjects who received placebo, but who were told they were receiving steroids.

Weightlifters need to develop large muscles.

Is it harmful to take steroids?
The poster shows some of the known side effects of steroids.

Questions

1. What is the effect on a male of removing the testes?
2. What conclusions can be drawn from the results of the 1996 study on testosterone about the effectiveness of using steroids in training?
3. What is the natural function of testosterone?
4. Give two ways in which using steroids enhances athletic performance.
5. Give two ways in which using steroids may affect health in women.
6. Explain what is meant by the 'placebo effect'.
7. Suggest a reason for the ban on the use of steroids.
8. Imagine you are an athletics coach. One of your athletes says that she knows that several of her competitors are taking steroids. She does not want them to out-perform her. What advice would you give her?

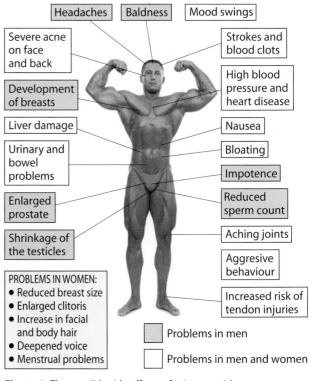
Figure 1 The possible side effects of using steroids.

ISA practice: testing hand-washes

Scientists are investigating the best hand-wash to use in a hospital. Your task is to do an investigation to compare the effect of antiseptic-based hand-washes and soap-based hand-washes on the growth of bacterial colonies.

Hypothesis

There is a link between the type of antibacterial substance used and the growth of bacteria.

Section 1

1 In this investigation you will need to control some of the variables.

   (a) Name one variable you will need to control in this investigation. *(1 mark)*

   (b) Describe briefly how you would carry out a preliminary investigation to find a suitable value to use for this variable. Explain how the results will help you decide on the best value for this variable. *(2 marks)*

2 Describe how you are going to do your investigation. You should include:

   - the equipment that you would use
   - how you would use the equipment
   - the measurements that you would make
   - a risk assessment
   - how you would make it a fair test.

   You may include a labelled diagram to help you to explain your method.

   In this question you will be assessed on using good English, organising information clearly and using specialist terms where appropriate. *(9 marks)*

3 Design a table that will contain all the data that you are going to record during your investigation. *(2 marks)*

   Total for Section 1: 14 marks

Section 2

A group of students, Study Group 1, investigated how effective two different hand-washes were in killing bacteria. They decided to place a drop of each hand-wash onto an agar plate and measured the radius of the clear zone that appeared. They repeated their results three times. Figure 1 shows the results they obtained.

Soap-based hand-wash	Antiseptic-based hand-wash
Radius of the clear zone in cm	Radius of the clear zone in cm
Plate 1	Plate 1
disc 1 0.8, disc 2 0.4, disc 3 0.6	disc 1 1.6, disc 2 1.6, disc 3 1.2
Plate 2	Plate 2
disc 1 0.8, disc 2 0.8, disc 3 0.8	disc 1 1.4, disc 2 1.8, disc 3 1.8
Plate 3	Plate 3
disc 1 0.6, disc 2 0.6, disc 3 0.8	disc 1 1.2, disc 2 1.4, disc 3 1.4

Figure 1 Results from Study Group 1's investigation.

4 **(a)** **(i)** What is the independent variable in this investigation?

   (ii) What is the dependent variable in this investigation?

   (iii) Name one control variable this investigation. *(3 marks)*

   (b) Plot a graph to show the link between the type of hand-wash used and the radius of the clear zone. *(4 marks)*

   (c) Do the results support the hypothesis? Explain your answer. *(3 marks)*

Below are the results of two other study groups.

Table 1 shows the results of another two students, Study Group 2.

Table 1 Results from Study Group 2.

	Mean radius of clear zone/cm		
	Plate 1	Plate 2	Plate 3
Soap-based hand-wash	0.4	0.6	0.5
Antiseptic-based hand-wash	1.5	1.7	1.2

Study Group 3 is a group of scientists who also investigated the effectiveness of a range of different hand-washes. The scientists investigated how well the 14 different hand-hygiene methods A–N, shown in Table 2, worked.

For their investigation the scientists recruited 70 volunteers. The volunteers were asked to wash their hands only with non-antimicrobial hand soap for seven days before the investigation. Five volunteers used each of the methods A–N during the investigation.

Each volunteer:

- washed hands with non-antimicrobial soap
- spread a standard suspension of a red-coloured bacterium over the hands for 45 seconds
- air-dried the hands for 60 seconds.
- used one of the hand hygiene methods, A–N, for 10 seconds
- rinsed the hands for 10 seconds.

Table 2 Hand hygiene methods used by Study Group 3.

Method	Active ingredient	Form	Method of application
A	60% ethyl alcohol	gel	waterless hand-rub
B	61% ethyl alcohol	lotion	waterless hand-rub
C	61% ethyl alcohol and 1% CHG	lotion	waterless hand-rub
D	62% ethyl alcohol	foam	waterless hand-rub
E	70% ethyl alcohol and 0.005% silver iodide	gel	waterless hand-rub
F	0.5% parachlorometaxylenol and 40% SD alcohol	wipe 256 cm^2	waterless hand-wipe
G	0.4% benzalkonium chloride	wipe, 296 cm^2	waterless hand-wipe
H	0.75% CHG	liquid	hand-wash
I	2% CHG	liquid	hand-wash
J	4% CHG	liquid	hand-wash
K	1% triclosan	liquid	hand-wash
L	0.2% benzethonium chloride	liquid	hand-wash
M	non-antimicrobial soap	liquid	hand-wash
N	tap water	liquid	hand-wash

The scientists from Study Group 3 carried out a second investigation using similar techniques, but this time using a virus instead of a bacterium.

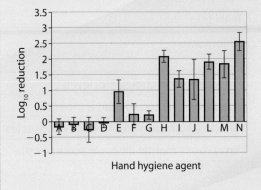

Figure 3 Results from Study Group 3's second investigation.

Each volunteer's hands were then sampled for the red-coloured bacterium by:

- placing each hand into a large latex glove containing 75 cm^3 of a sterile sampling solution
- having the glove massaged for 30 seconds.

5 cm^3 of the sampling solution was then spread over a nutrient agar plate. The agar plates were incubated at 25 °C for 24 hours, after which the number of colonies of the red-coloured bacterium was counted. Each colony had developed from a single bacterium.

Figure 2 shows some of the scientists' results.

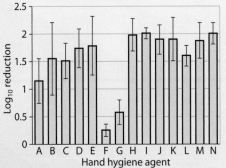

Figure 2 Some of the scientists' results.

A log reduction of 1 means that there are 10 times fewer bacteria in the sample.

A log reduction of 2 means that there are 100 times fewer bacteria in the sample.

A log reduction of 3 means that there are 1000 times fewer bacteria in the sample.

The line associated with each bar on the chart shows the range of results for each hand hygiene method.

5 Describe one way in which Study Group 2's results are similar to or different from the results from Study Group 1, and give one reason why the results are similar or different. *(3 marks)*

6 **(a)** Summarise briefly the results from Study Group 3. *(3 marks)*

(b) Does the data from the study groups support the hypothesis being investigated? To gain full marks you should use all of the relevant data from Study Groups 1, 2 and 3 to explain whether or not the data supports the hypothesis. *(3 marks)*

(c) The data from the other groups only gives a limited amount of information. What other information or data would you need in order to be more certain as to whether or not the hypothesis is correct? Explain the reason for your answer. *(3 marks)*

(d) Use the results from Study Groups 1, 2 and 3 to answer this question. What is the relationship between the type of hand-wash and the effect on bacteria? How well does the data support your answer? *(3 marks)*

7 Look back at Study Group 1's method. If you could repeat the investigation, suggest one change that you would make to the method, and give a reason for the change. *(3 marks)*

8 Suggest how ideas from your investigation and the scientists' investigations could be used to advise hospitals about the type of hand-washes to use. *(3 marks)*

Total for Section 2: 31 marks

Total for the ISA: 45 marks

Assess yourself questions

1 Match drugs A, B and C with the correct statements in Table 1.

   A   cannabis

   B   steroid

   C   thalidomide

Table 1

Information	Drug A, B or C
reduces blood cholesterol levels	
was recently used to treat leprosy	
used by some athletes to build up muscles	
may cause mental illness in some people	

(3 marks)

2 The drug thalidomide was once banned.

Now the drug is being tested to see whether it can be used to treat the disease AIDS.

Match words A, B, C and D, with the numbers 1–4 in the sentences.

   A   the government

   B   pregnant women

   C   research scientists

   D   volunteers

The trials will be carried out by . . . 1 . . . .

In the trials, the drug will be given to . . . 2 . . . .

The drug should not be given to . . . 3 . . . .

The final decision on whether the drug is licensed for use by AIDS patients will be taken by . . . 4 . . . .   *(4 marks)*

3 Caffeine is a drug present in many drinks.

  (a) Explain what is meant by 'drug'.   *(1 mark)*

Scientists asked 2500 pregnant women to record how much tea, coffee and other foods that contain caffeine they consumed each day. The scientists then compared the results of the survey with the birth mass of the women's babies.

They found that women who consumed more than 200 mg of caffeine per day were more likely to give birth to smaller babies

Following the research, the Food Standards Agency lowered the recommended caffeine intake for pregnant women from 300 mg a day to 200 mg.

  (b) Are the results of this survey likely to be reliable? Explain the reason for your answer.   *(1 mark)*

  (c) Figure 1 shows the amount of caffeine in different drinks

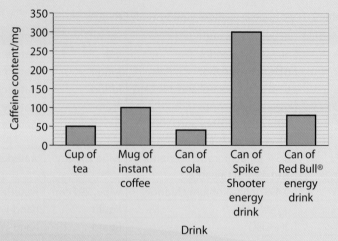

Figure 1 Caffeine concentration.

   (i) What is the maximum number of cups of tea a pregnant woman could drink in one day and still be within the new recommended limit?   *(2 marks)*

   (ii) Caffeine increases heart rate and is linked to high blood pressure. Using the data in the chart, what type of drink would you recommend for a person whilst at work in an office? Explain the reason for your answer.   *(2 marks)*

4 Many people become addicted to drugs.

  (a) Explain how a person becomes addicted to drugs.   *(2 marks)*

Figure 2 shows changes in the number of drug-related deaths in the UK between 1993 and 2007.

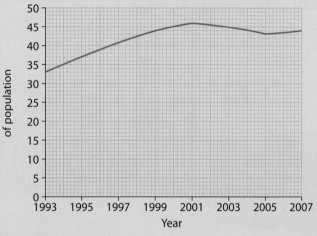

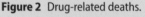

Figure 2 Drug-related deaths.

  (b) **(i)** Describe the pattern shown by the data.   *(2 marks)*

     (ii) Suggest explanations for the changes in the number of deaths.   *(2 marks)*

  (c) A city has 2 million inhabitants. How many drug-related deaths would be expected in that city in 2007?   *(2 marks)*

5 New drugs have to be extensively tested and trialled before they can be given to patients.

List A gives three stages in drug testing.

List B gives information about each stage.

Draw one line from each stage in List A to information about the stage in List B.

List A	List B
Stage	Information
trials that include a placebo	testing to see if the drug is toxic
trials on small group of volunteers	testing to detect side effects
laboratory trials	tests in which neither the patient nor doctor knows who has been given the drug

(3 marks)

6 New drugs must be tested before use. A form of ultrasound is being used by scientists to test the effectiveness of drugs designed to break down potentially life-threatening blood clots. Scientists from King's College of Medicine in London claim the technique provides a more reliable measure of the effectiveness of drugs than was previously available, and could remove the need to test new drugs on animals.

They have used the technique to test the effectiveness of a new drug – GSNQ – which dissolves blood clots. This reduces the risk of strokes. GSNQ was compared with the standard treatment of aspirin and heparin in a group of 24 patients who underwent surgery to clean a major blood vessel in the neck. Patients treated with GSNQ were found to have significantly lower numbers of clots during a three-hour period after the operation.

A member of the research team said: 'Before this technique assessing a drug meant either doing animal tests, or taking blood from people and studying it under the microscope. Neither was a very good measure of what would actually happen when the drug was used in people.'

New drugs will still have to be thoroughly assessed in large-scale clinical trials, but the new technique will help scientists to decide which products should go to a full trial.

(a) How did the scientists measure the effectiveness of GSNQ? *(1 mark)*

(b) Give three advantages of the above method of testing GSNQ over traditional drug-testing methods. *(3 marks)*

(c) Explain why GSNQ will still need to be assessed in large-scale clinical trials before it is approved. *(2 marks)*

7 Table 2 Comparing the effects of a number of drugs (3 = highest effect).

Drug	Mean dependence	Mean physical harm	Pleasure	Social harm	Healthcare costs
heroin	3.00	2.78	3.0	3.0	3.0
cocaine	2.39	2.33	3.0	2.5	2.3
barbiturates	2.01	2.23	2.0	1.9	1.7
street methadone	2.08	1.86	1.8	1.9	2.0
LSD	1.51	0.99	1.9	1.3	1.5
ecstasy	1.23	1.13	2.2	1.3	1.1
cannabis	1.13	1.05	1.5	1.0	1.1
anabolic steroids	0.88	1.45	1.1	0.8	1.3

Describe the conclusions that can be drawn from this data. *(5 marks)*

8 Scientists tested the drug thalidomide for effectiveness as a sleeping pill.

They gave tablets to two groups, each of ten volunteers, X and Y.

The tablets given to group X contained thalidomide but the tablets given to group Y did not. Neither group knew which type of tablet they were taking.

The scientists observed the time taken for each person to fall asleep before and after taking the tablets.

The test was repeated a further two times and the mean time taken to fall asleep calculated.

Table 3 The results of the test.

Group	Mean time taken to fall asleep/minutes										
	Person	1	2	3	4	5	6	7	8	9	10
X	Before taking the tablets	47	39	52	37	40	32	30	28	46	48
	After taking the tablets	27	32	19	17	24	36	14	22	29	31
	Person	1	2	3	4	5	6	7	8	9	10
Y	Before taking the tablets	36	48	52	36	44	34	34	26	44	46
	After taking the tablets	36	49	48	33	45	35	28	24	40	41

(a) Give three ways in which the scientists tried to get reliable results. *(3 marks)*

(b) Summarise the results for
   (i) Group X *(2 marks)*
   (ii) Group Y *(2 marks)*

(c) Suggest an explanation for the results of person 10 taking tablet Y. *(1 mark)*

Here are three students' answers to the following question:

Read the information about chickenpox.

Chickenpox is a common, highly infectious disease caused by a virus. Sufferers usually experience a blistering rash accompanied by mild fever, loss of appetite, headache and sore throat. Cases of the disease are not recorded in the UK, but they are thought to number about 600 000 a year, and by the age of 12 some 90% of us will have had it. Children with chickenpox are barred from nursery or school.

Researchers found that in one year in the UK, 112 children had serious complications of chickenpox. Six of these children died. Of these, four had a chronic health problem, such as HIV or cerebral palsy. When compared with other illnesses of childhood, it is less serious than measles, but potentially more problematic than mumps. This is because after the initial infection the virus remains dormant in the central nervous system for the rest of an individual's life and can reappear later as a painful condition called shingles.

All children in the USA are vaccinated against chickenpox. Researchers there have found that vaccinating children has raised the rate of shingles among the elderly by 90%. This is because once a person has had chickenpox, every time they come into contact with an infected child their body's defence system builds up immunity against the disease. If children no longer harbour the disease, a rise in the number of shingles cases in the elderly follows.

A UK government committee is considering whether all children in the UK should be vaccinated against chickenpox.

Using the above information, evaluate the case for vaccinating all UK children against chickenpox.

In this question you will be assessed on using good English, organising information clearly and using specialist terms where appropriate. (6 marks)

'Read the three answers below, together with the examiner comments. Then check what you have learned and put it into practice in any further questions you answer.'

B Grade answer

Student 1

> Always use the correct term, in this case vaccination; 'jab' is a colloquialism and should be avoided.

> The answer should not start with a conclusion.

> No credit is gained for copying large amounts of text from the passage.

I think that all UK children should be given a chickenpox jab because there are over 600 000 cases per year. It is a very unpleasant disease in which children experience a blistering rash accompanied by mild fever, loss of appetite, headache and sore throat. Also 112 children got serious complications and six children died. It can also cause shingles in the elderly.

Examiner comment

This candidate stated three reasons for vaccination against chickenpox, but has given no reasons against vaccination so would be limited to two of the five marks. In questions of this type, two marks are allotted for 'pros', two marks for 'cons' and one mark for a conclusion. The conclusion mark is not awarded if the candidate has not given both 'pros' and 'cons'.

(A) Grade answer

Student 2

It is much better to keep 'pros' and 'cons' separate rather than mixing together in one sentence.

It is not fully clear whether this is a 'pro' or a 'con'.

Vaccination would reduce the number of cases of chickenpox, but chickenpox is not a very serious complaint so it might be a waste of money. Also, parents have to take time off work to look after children with the disease. If childhood vaccination is introduced, more elderly people may get shingles. Overall I think that the case against vaccination outweighs the case for vaccination.

No marks are given for a conclusion that does not give a reason and that does not refer to both pros and cons.

Examiner comment

Although the answer has pros and cons jumbled together the candidate did give two reasons in favour of vaccination: the reduction in the number of cases and the fact that parents usually have to take time off work to look after children. The candidate also gave two reasons against vaccination: the fact that it is not usually a serious complaint and that it might increase cases of shingles in the elderly. The candidate should have continued the last sentence with the word 'because' followed by a reason to get the conclusion mark.

(A*) Grade answer

Student 3

The candidate makes it quite clear that they are talking about 'pros'.

The candidate now makes it quite clear that they are moving onto 'cons'.

The case for vaccination is mainly the fact that 150 children a year get serious complications from the disease and that the virus can stay dormant in the body and reappear as shingles in later life. Vaccination would prevent both of these.

The case against vaccination is that the disease is very mild in the majority of cases, so vaccination might not be cost-effective. There is also the likelihood that vaccinating all children will result in an increase in the number of cases of shingles in the elderly.

Overall I think that vaccination should not be given to all children because the increase in the number of cases of shingles in the elderly is far more serious than the mild symptoms experienced by the majority of children who get chickenpox.

The candidate now starts a third paragraph for the conclusion and uses the word 'because' to explain the reason for their decision.

Examiner comment

A well laid out answer that directs the examiner to at least two 'pros', at least two 'cons' and a reasoned conclusion. The candidate has avoided copying out large tracts from the passage.

The conclusion refers to both a 'pro' and a 'con' then goes on to differentiate between the effect of the two.

- When answering evaluation questions, first read the passage and mark up 'pros' and 'cons' as you go through.
- Start your answer with a paragraph containing at least two pros.
- Write a second paragraph including at least two 'cons'.
- Write a conclusion that includes the word 'because' so that you give a reason for your conclusion.
- Refer to at least one 'pro' and at least one 'con' in your conclusion.

Environment and evolution

The world is filled with a breathtaking array of plants, animals and microorganisms, many of which have distinct and obvious adaptations to the physical conditions in which they live and to the organisms that surround them. This section begins with opportunities to explore some of those adaptations and to see how changes, such as those of climate and of pollution, affect organisms and their distribution. It continues with an exploration of the role of microorganisms in the natural cycles of decay and the carbon cycle.

The next chapter introduces the concepts of variation and genes, to provide an understanding from which to explore the developments of genetic engineering and cloning. These techniques offer many opportunities for the future, but have the potential to cause problems as well as raise ethical issues that need to be discussed before decisions can be made.

The final chapter brings together the elements of adaptation and genetic variation in the topics of evolution and natural selection. A discussion of how Darwin's theory of evolution developed provides a framework within which to explore the development of scientific theory in general, and the role of the scientific community in that.

Test yourself

1 What causes variation in living organisms?
2 How do organisms inherit characteristics from their parents?
3 How are organisms affected by the environment and by each other?
4 How is human activity affecting the environment?
5 How do we develop scientific ideas from experiment and investigation?

Objectives

By the end of this unit you should be able to:

- describe a range of adaptations in plants, animals and microorganisms to the environment in which they live
- interpret evidence of changes in the distribution of organisms in terms of changes in their environment
- explain the critical role of microorganisms in nutrient recycling and the carbon cycle
- describe the roles of environment and genes in creating variation
- describe the development of new techniques in cloning and genetic engineering and explain why they are being developed
- make informed judgements about economic, social and ethical issues about cloning and genetic engineering
- explain how natural selection can lead to the evolution of new species
- using evolutionary theory as an example, explain how scientific ideas develop into theories.

Plant adaptations

Learning objectives

- describe some features (adaptations) of plants that help them grow well in different conditions
- describe specific adaptations that help plants survive in dry environments
- explain how different adaptations to the environment help plants to survive.

The leaves and branches of coniferous trees allow snow to slide off quickly without damage.

Science skills

a How would you investigate the effect of different amounts of shade, or different levels of nutrients, on the growth of plants?

Examiner feedback

The conservation of water by plants living in dry conditions and the efficient uptake of water when water is available, through adaptations in the root systems, are ideas that are often questioned in exams.

Adapting to the environment

All organisms need food, water and **nutrients** to grow. They get these from the environment where they live. Different environments have different **physical conditions**, such as temperature, and amounts of light, water and nutrients. To grow well and produce offspring, an organism needs particular features that help it get what it needs from its environment. We call these features **adaptations**.

Growing in the cold

Where it is very cold, such as in the Arctic tundra, few plants can grow because ice forming inside cells can damage them. These plants often have a rounded shape that helps to **insulate** the inner parts and keep them warmer.

In coniferous forests of the northern hemisphere, the temperature is a little warmer, but heavy snow is a danger as it can break branches if it piles up.

Growing in rainforests

In a tropical rainforest, the temperature may be good for growth but rainfall can be so heavy that it damages leaves. Many leaves have shiny surfaces and pointed tips to the leaves to help the rain run off quickly, or are divided into many sections so water can pass through easily. At the top of the trees, there is plenty of light for **photosynthesis**, but the forest is usually so dense that plants growing near the forest floor get little light.

Many plants can grow in a tropical rainforest where it is warm all year round but they need adaptations for other conditions.

The red backs of the leaves help ground-level plants capture more light for photosynthesis.

Growing in the dry

Plants, like cacti, that grow in places where it is dry have very wide root systems to collect as much water as possible when it rains, or very deep root systems that can tap into water far underground. They may also have thick leaves or a thick body that contains tissue for storing water that the plant can use when there is a **drought**.

Tiny holes, called **stomata**, in the surfaces of leaves let in carbon dioxide for photosynthesis, but these holes also let water out. Plants in dry areas must reduce water loss through stomata, so they may have no leaves at all. A few stomata in the stem surface are protected by hairs or by being placed deep in ridges. This reduces the speed of air moving across the stomata and so reduces the rate of evaporation. The plant can continue to photosynthesise because the stems are green.

The effect of soil

The amount of nutrients in the soil will affect plants. Where the soil contains many nutrients, many different kinds of plants may be able to grow. However, few plants grow well on poor acid soil because it contains so few nutrients. For example, sundew plants get extra nutrients by trapping insects on their leaves and digesting them.

The hairy surface of the cactus reduces the speed of air flow across the stomata and so reduces the rate of evaporation of water from the plant.

The sundew plant lives in waterlogged and nutrient-poor soil.

Examiner feedback

It is important not just to describe the adaptations of an organism to particular conditions in the environment, but also to explain why those adaptations increase the chances of survival.

Questions

1   Describe one adaptation of a plant to extreme cold, and explain how this helps the plant to survive.

2   Describe three adaptations of rainforest plants to their environment and explain how they help plants to survive.

3   A rainforest leaf that is split into sections can be bigger than one that isn't split. Explain why this is an advantage.

4   List as many ways as possible that plants which live in dry conditions can make sure they have enough water during long droughts, and explain the importance of each way.

5   In a tropical rainforest the tallest trees are over 60 m, below which are 'layers' of smaller trees and shrubs. Suggest the different adaptations you would expect to see in the tallest trees and the shrubs.

6   Marram grass grows on sand dunes on sea shores and has very deep roots. The leaves are curled, with the stomata tucked inside. Explain as fully as you can how these adaptations help the marram grass to survive.

7   Sundew plants grow slowly. Explain in detail why sundew plants are only found on poor acid soils.

8   The leaves of coniferous trees are thin, waxy needles that remain on the tree all year round. The stomata are hidden in pits in the needle surface. Explain in as much detail as you can why these and other adaptations of conifer trees help them dominate northern landscapes.

Science in action

High concentrations of metals in soil are usually poisonous to plants. However, a few plants are adapted to these conditions. These plants can be grown on the spoil heaps from mines where metals such as copper and lead have been extracted. The plants help to remove the metals from the soil, which makes it less poisonous for other plants.

Taking it further

At A level you will learn how plants change the environment in which they live, such as by creating soil around their roots or drying out wetlands. These changes create conditions that are better suited to other plants, which may change conditions in other ways. The gradual replacement of one group of species with others over time is called succession.

Animal adaptations

Science in action

Studying the hollow hairs of caribou, which live in the frozen north of Canada, showed that the air inside the hairs acts as good insulation from the cold. This led to the development of hollow polyester fibres for use in duvets and insulated clothing to keep us warm.

Desert hopping mice can usually only be found on the surface at dawn and dusk.

Examiner feedback

You will not be asked about Bergmann or Allen in an exam. However, you should be able to explain how the surface area to volume ratio of an animal has a direct effect on where it can live and why. Remember that a large animal has a large surface area, but a small surface area to volume ratio.

Staying warm

Where environmental conditions, such as temperature and water availability, are extreme, animals also need particular adaptations to survive and grow well.

Where it is very cold, animals need to reduce heat loss. Land mammals, such as polar bears and caribou, have very thick fur that insulates them from the air so that body heat is lost more slowly to

Before winter, the polar bear grows a thicker coat and eats lots of food. When there is no food, it can reduce its rate of energy use to conserve fat reserves.

the cold environment. Extra fat, in a layer under their skin, increases the insulation and can give the animal energy when food is scarce. During the summer, these animals **moult**, or shed the outer thick layer of fur so that it is easier to lose body heat. Sea mammals, such as seals, need a very thick layer of fat to insulate against heat loss to the water.

Keeping cool and avoiding thirst

In deserts, there is very little water for much of the time. It can also be very hot during the day and very cold at night.

Small desert animals, such as the hopping mice of Australia, live mostly in burrows where the temperature is more constant and not as extreme as above ground. They also do not need to drink water, as they get all the water they need from their food. Their kidneys **excrete** urine that contains very little water. Large desert animals, such as camels, can tolerate higher levels of dehydration than non-desert species.

Science skills

Table 1 The surface area and volume of different sizes of cubes.

Length of one side/cm	Surface area of cube/cm^2	Volume/cm^3
1	6	1
2	24	8
3	54	27
4	96	64

a Calculate the surface area to volume ratio for each cube.

b Heat is generated in an animal's body by the *volume* of tissue. Heat is lost by an animal from the *surface area* of its body. The larger the surface area to volume ratio, the faster the animal will lose body heat. Which size loses heat faster: the smallest or the largest?

Body size and shape rules

The 19th century biologist Carl Bergmann observed that birds and mammals of the same or similar species tend to be larger and heavier when they live in colder climates. 'Bergmann's rule' states that there is a correlation between body mass and average annual temperature. Bergmann's explanation for the correlation was that animals lose heat at the surface of their bodies so, if two animals are shaped identically, the temperature of the larger one will drop less rapidly.

Science skills

c How would you test the idea that penguins huddling in Antarctic winters is a behavioural adaptation to the cold?

Huddling penguins.

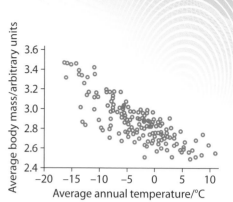

Figure 1 Body mass of many different animals compared with temperature where they live.

Joel Allen gathered data relating climate to variation in animals. Allen's rule states that the extremities of organisms (limbs, tails and ears) of animals that need to maintain their temperature, such as mammals, are longer in warm climates than in cold climates because they act as heat-radiating organs.

Questions

1   Describe the adaptations of the polar bear to life in the Arctic, and explain how they help it survive.

2   **(a)** Explain how small animals can survive life in the desert. **(b)** Suggest what adaptations large desert animals need to survive, and how these adaptations enable them to survive.

3   Use the information about surface area and volume of cubes to predict what impact body size in animals will have in: **(a)** cold climates **(b)** hot climates.

4   Do the data shown in the graph in Figure 1 support Bergmann's rule? Explain your answer.

5   Do the kit fox and Arctic fox comply with Allen's rule? Explain your answer.

6   Use Bergmann's rule to help explain why Emperor penguins huddle during polar winter storms.

7   During the Arctic winter, a mother polar bear stays in a den with her cubs. Explain fully why this is an adaptation to living in polar conditions.

8   Whale blubber is not just a layer of fat, but includes blood vessels that can be adjusted to allow more or less blood near to the skin. Explain in as much detail as you can why a whale needs such a complex system.

The kit fox lives in the hot dry regions of Mexico.

The Arctic fox lives in northern Canada.

Surviving the presence of others

The blue poison dart frog boldly sits in full view of potential predators.

Bluebells cannot compete with trees in leaf for the light they need.

Keeping others away

The environment of an organism includes not only the physical conditions but also other plants and animals. If an organism is to grow, mature and produce offspring, it must avoid being eaten by other organisms.

Many organisms have special adaptations to avoid being eaten. Many plants, for example, cacti, have physical deterrents such as thick spines to deter herbivores. Other plants, like ragwort, use chemical deterrents such as poisons that taste unpleasant or can kill.

Many animals also use poison to deter predators. They advertise their poison with bright colours. Other animals hide from predators using **camouflage**. The colours and patterns on their bodies make them much more difficult to see against their usual background.

Science skills
a European banded snails come in a wide variety of shell colours and patterns. How would you test the idea that the variety is the result of camouflage against predation by birds?

Competing for resources

It is rare that there is so much of a resource that all organisms can get what they need. When resources are limited, organisms have to **compete** with each other. Those organisms that are most successful in getting the resource will grow better and produce more offspring.

Plants compete with each other for water, light and nutrients if they are growing close together. Some ground-living plants in woodland, such as bluebells, are adapted to grow, flower and set seed before the trees have fully grown their leaves and block out the Sun.

Animals of different species may compete with each other for the same food. For example, a leopard may kill an antelope to eat, but there will be many other animals, including lions and hyenas, that could take the antelope from the leopard. Leopards often take their kills up into a tree to keep them from other animals.

Animals of the same species may also compete with each other for resources other than food. In species where males mate with many females, competition to father the next generation can be intense.

Many birds have **territories** when they are breeding. They keep other birds out of it to give them enough space to rear their young. Robins need a territory that is large enough to provide all the food they need for their young. However, birds that feed away from the nest site, such as gannets or penguins, will have small nesting territories, just large enough to be out of pecking distance of other adults.

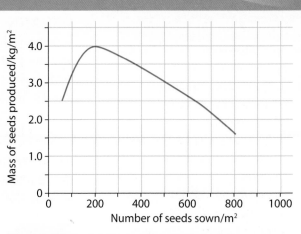

Figure 1 Graph of yield versus density at planting.

Questions

1 Using the graph in Figure 1, describe how the seed yield of a crop depends on the density of plants in the crop.

2 Suggest why polar bears are white.

3 The robin is well known for its red breast and for singing loudly in spring. **(a)** What are the disadvantages of these adaptations? **(b)** What are the advantages?

4 **(a)** Copy the table headings from below and complete your table to show all the resources that animals might compete for. **(b)** Give examples of animals for each resource. **(c)** Explain how each animal is adapted to compete for that resource.

Resource	Example	Adaptation

5 Explain why blue poison dart frogs are blue as well as poisonous.

6 **(a)** It takes a lot of energy for a male peacock to grow his tail feathers and display them. Why does he bother? **(b)** Suggest an advantage to the females (in terms of genes) who mate with the male giving the best display.

7 Annual poppies (plants that live only one year) often suddenly appear on land that has recently been cleared of plants, but not in the following year. Explain this observation as fully as you can in terms of competition.

8 Using examples from what you have learnt so far, describe conditions in which plants may compete for light, water and nutrients.

9 Another strategy for avoiding predation is to look like something dangerous: for example, hoverflies look like wasps, but have no sting. Evaluate the advantages and disadvantages to the hoverfly and wasp of this similarity.

A*

Male peacocks display their huge tail feathers and call when there are females around. The male with the best display will mate with most females.

Extreme microorganisms

Learning objectives

- describe some extreme conditions that microorganisms can live in
- analyse the adaptations that microorganisms need to survive in some extreme conditions
- explain how bacteria that live in extreme conditions are important for other organisms.

Examiner feedback

Be careful not to write that all extremophiles are microorganisms. A few animals are extremophiles, such as the Pompeii worm, which lives in extremely hot water close to deep sea vents (below), and tardigrades (microscopic animals often called water bears), which can survive temperatures as high as 151°C and as low as −270°C.

All the organisms living around this hydrothermal vent deep in the ocean depend on chemosynthetic bacteria.

Extreme temperatures

Organisms that can survive in the most extreme conditions are called **extremophiles**, and they are mostly microorganisms. In such conditions, there are few other organisms to compete with, but the adaptations needed to survive are extreme too.

The single-celled alga *Chlamydomonas nivalis* contains an extra-red pigment that protects it from the intensely bright light in snowy places that would destroy other cells. The colour also absorbs radiation, helping to keep the cells warmer than the snow around them. After the winter, when the cells have been buried by more snow, they grow **flagella** that help them move back up to the surface.

The temperature of the water here is over 60 °C, but there are many kinds of bacteria living in the water and around the edges of the hot spring.

Extreme high temperatures can be found in a few places on Earth, such as around hot springs, or around **hydrothermal vents** deep in the ocean. The heat comes from the rocks below ground. Most organisms die above about 40 °C because the proteins in their cells break down. Bacteria that survive at higher temperatures have proteins that don't break down as easily.

No light

Deep in the ocean, water pressure is great and there is no light. Amazingly, there are large communities of organisms that can live around hydrothermal vents on the ocean floor. They all depend on bacteria that use **chemosynthesis**, which combine chemicals from the water using heat to make sugars for food. These bacteria are the **producers** in these communities, like green plants on the Earth's surface.

Too many nutrients

Organisms need nutrients to make chemicals inside their cells. If the concentration of nutrients outside the cells gets too high, it draws water out of the cells. For most organisms this is fatal. Some microorganisms, however, have high levels of other chemicals, such as **amino acids** or sugars, inside their cells that stop the water moving out.

This lake is so salty that the man floats. This concentration of salt in his body would kill him, while the bacteria that make the water pink thrive in it.

Too little oxygen

Most large organisms depend on oxygen for **aerobic respiration**, but in a few places there is very little or no oxygen, such as deep below ground. Here there are microorganisms that are adapted to get their energy in other ways. Some use chemicals, such as sulfur, instead of oxygen to release energy in **anaerobic respiration**. Another example of anaerobic respiration is the **fermentation** of glucose by yeast.

Science in action

Soy sauce is made by fermenting soy beans. The fermentation needs microorganisms to break down the proteins in the beans. The strains of microorganisms used for fermentation must also tolerate high salt levels.

Many kinds of bacteria live in conditions of low oxygen inside the cow's gut and help to digest the grass. Without them, the cow could not survive on a diet of grass.

Questions

1. List the advantages and disadvantages of being an extremophile.

2. List all the extreme conditions described on these pages, and describe adaptations that make survival possible in those conditions.

3. The ability of microorganisms to live in extreme conditions makes life possible for many other organisms. Explain what this statement means.

4. Two competing theories for the origin of life on Earth are that:
   - bacteria from other planets came to Earth on meteorites
   - bacteria first originated around hydrothermal vents because the surface of the Earth was too extreme.

   (a) Explain why both these theories are possible. **(b)** What else would you need to know in order to evaluate which theory was the more likely?

5. Suggest why microorganisms are more able to live in extreme conditions than larger organisms.

6. A student had some onions. She filled two jars with onions and added pickling vinegar. After six months, the onions she hadn't pickled had gone mushy, but the onions in one jar were still crunchy. After two years she opened the second jar and the onions were mushy. Explain these observations.

7. There are an estimated 10 times more bacterial cells in your body than human cells. Most of these bacterial cells live in your gut and play an important role in digestion. Evaluate the advantages and disadvantages of using antibiotics for curing infections.

Science in action

In November 1969, astronauts brought a camera back from the Moon that had been there since April 1967. Live bacteria were found growing inside the camera. Having contaminated the camera on Earth, these **Streptococcus mitis** bacteria survived 31 months in the vacuum of the moon's atmosphere.

Taking it further

Without the microorganisms in your gut you would be very ill, and might not even be able to live. They not only help you digest your food, they also produce vitamins that you need to stay healthy, but that aren't found in food. They also help 'train' your immune system to recognise what is foreign and should be attacked, and what is safe and should be left alone.

Pollution indicators

Pollution changes the environment

Pollution from human activity damages the environment and the organisms that live there. Many industrial processes burn fossil fuels for energy or to release other chemicals for manufacture. Some of the waste gases from this burning are poisonous, or are acidic; when they dissolve in water droplets in cloud, they can form **acid rain**. The acid is damaging to many plants and animals.

We also pollute water by releasing chemicals directly into it, for example when too much fertiliser is put on farmland or when chemicals are released from factories or sewage treatment plants into the water. There are laws against causing air and water pollution, but they can be difficult to monitor and enforce.

Science skills

The lower the pH, the more acidic a solution is.

a Describe the relationship between the number of species of plants and the acidity of the water in a pond.

b Give two explanations for the relationship between number of species of insect larvae and water acidity.

c How could you improve the reliability of your answers?

Table 1 Pond data.

Pond	pH of water	Number of species of insect larvae	Number of species of plants
1	4.4	4	8
2	4.8	5	11
3	5.7	9	16
4	6.6	19	23
5	8.1	14	21

If there is a high level of pollution, organisms may be killed and the distribution of species will change. We can use the presence of organisms that are particularly affected by, or tolerant of, pollution as **pollution indicators**.

Lichens and air pollution

Lichens are often found on trees and walls. There are many different species; some can only grow where there is no air pollution while others are tolerant of different kinds of air pollution.

Surveys of lichens can show quickly and easily how polluted the air has become since the lichens started growing. Recent surveys indicate that sulfur-intolerant species are colonising city areas where they had died out, showing that sulfur dioxide pollution in these areas has decreased.

The lichen on the left is a clean-air species. The lichen on the right tolerates some air pollution.

Practical

Carry out a lichen survey in your area, and use the species you find to decide where there is the most air pollution.

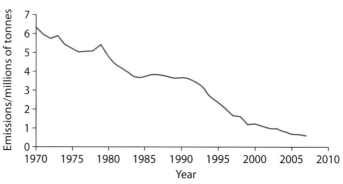

Figure 1 Sulfur dioxide emissions into air in the UK have reduced as we changed to low-sulfur fuels, and because of laws that control emissions from industry.

Aquatic invertebrates as pollution indicators

Most **aquatic** organisms get the oxygen they need for respiration directly from the water they live in, not from air. Fertilisers and sewage contain high levels of nitrogen and phosphate. If they drain into water, they can cause plants, algae and bacteria to grow rapidly and use up the oxygen during respiration.

As oxygen concentration drops, some organisms, such as mayfly nymphs, die because they can't get the oxygen they need. Other organisms can survive in water that has a very low oxygen concentration because they have special adaptations. The bloodworm is red because it contains haemoglobin, which combines with oxygen, as it does in our red blood cells. Different aquatic invertebrates can tolerate different oxygen concentrations, so we can use their presence or absence to indicate how polluted the water is.

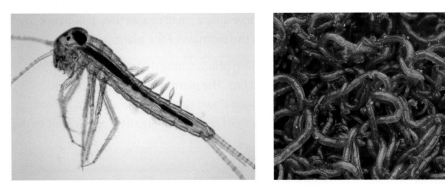

The mayfly nymph (left) can only live in unpolluted water, but the bloodworm (right) can survive in highly polluted water.

Questions

1   Define the word *pollution* and give examples of air and water pollution.

2   Explain why adding nitrogen to water reduces the oxygen concentration of the water.

3   In the 1970s it was shown that forests in Scandinavia, far from any industrial areas, were being damaged by acid rain. Suggest how this could happen.

4   Explain how bloodworms are adapted to low oxygen concentration.

5   Use the graph in Figure 1 to suggest what you would expect to see from lichen surveys: **(a)** from the Scottish Highlands in 1970 and today **(b)** in the centre of Manchester in 1970 and today. Explain your answers.

6   Compare the use of lichen surveys with roadside air pollution monitoring, and suggest the advantages and disadvantages of each.

7   Explain in detail why it may be easy to identify the source of water pollution caused by a poisonous chemical but difficult to find the source for high levels of nitrate in water.

Energy in biomass

About biomass and energy

The amount of 'stuff' that makes up your body is your **biomass**. Your biomass increases when food you eat is used to make more cells and tissues. If we measured the mass of all that food, we would have the biomass of your food.

We can also think in terms of energy. The energy in biomass is in the form of **chemical energy**, stored in the bonds and structures of the chemicals that make up cells.

Making plant biomass

The energy for making more plant biomass comes from the Sun. During photosynthesis, light energy is used to join water and carbon dioxide to make the carbohydrate glucose. Although leaves are adapted for capturing as much sunlight as possible, Figure 1 shows how only a small proportion of light energy is transferred into chemical energy in the biomass of the plant.

Some of the carbohydrates made during photosynthesis are converted into other chemicals, such as proteins and fats. The plant needs energy to build these, so other carbohydrates are broken down during **respiration** to release energy. However, some of the energy from respiration also escapes as heat to the environment.

From plant biomass to animal biomass

When an animal eats, some of the chemicals from its food are absorbed into its body, the rest are egested (expelled from the body) as **faeces**. Absorbed food chemicals may be built up into fats, carbohydrates and proteins and other chemicals that make the cells. These reactions need energy. Energy comes from breaking down carbohydrates during respiration. Respiration also supplies the energy for all the other living processes including movement. As with plants, some of this energy from respiration escapes to the environment as heat. The animal also makes waste products that still contain chemical energy, which it **excretes** into the environment in urine.

Biomass in food chains

Since energy is lost to the environment at each stage of a food chain, there is less biomass in each level of the chain. We can represent this in a **pyramid of biomass**, which shows the amount of biomass at each level of the food chain drawn to scale.

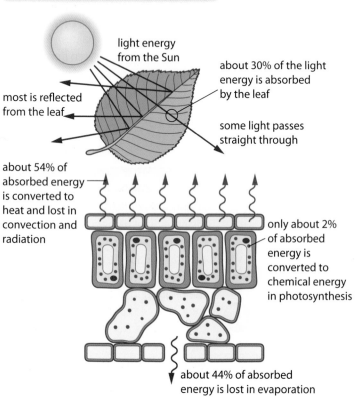

light energy from the Sun

about 30% of the light energy is absorbed by the leaf

most is reflected from the leaf

some light passes straight through

about 54% of absorbed energy is converted to heat and lost in convection and radiation

only about 2% of absorbed energy is converted to chemical energy in photosynthesis

about 44% of absorbed energy is lost in evaporation

Figure 1 What happens to light energy falling on a leaf.

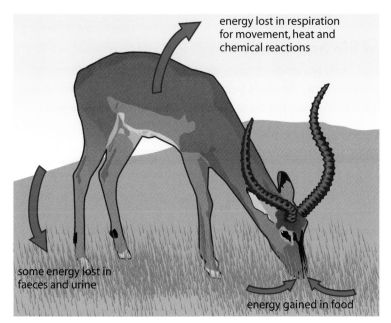

Figure 2 Energy losses from an animal.

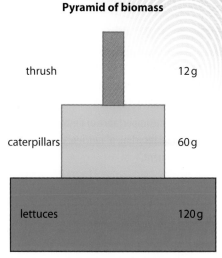

Figure 3 A pyramid of biomass for the food chain lettuce → caterpillar → thrush.

Questions

1 Describe the adaptations of leaves for collecting light energy.

2 **(a)** Identify all the forms of energy shown in the diagram of the leaf in Figure 1. **(b)** Starting with a block that represents all the light energy from the Sun that reaches a leaf, sketch a diagram that shows what happens to all of that energy.

3 Sketch the energy gains, losses and stores of a carnivore, such as a lion.

4 Draw a food chain that includes you and something that you ate today. Identify all the changes in forms of energy through that food chain.

5 On graph paper, draw a pyramid of biomass for the following food chain from the ocean: producers (phytoplankton) 50 g/m^2, primary consumers (krill) 20 g/m^2, secondary consumers (penguins, minke whales, crab-eating seals) 5 g/m^2, tertiary consumers (killer whales, leopard seals) $<1 \text{ g/m}^2$.

6 **Table 1** The percentage of energy taken into an animal's body that is used for making new tissue.

Group	Percentage energy transfer into tissue
ectothermic ('cold-blooded') vertebrates, e.g. reptiles, amphibians	~10%
medium–large **homeothermic** ('warm-blooded') vertebrates, i.e. mammals and birds	1–2%
small homeothermic vertebrates, e.g. wren, shrew	~0.4%

Suggest explanations for the differences between the animal groups.

7 Explain fully how you would evaluate the reliability of data shown in your pyramid of biomass for Question 5.

8 Humans currently use over one-third of all useful plant material that grows on Earth every year. Explain fully the impact of an increasing human population on other life in terms of energy flow.

Science skills

Gathering reliable data to make a pyramid of biomass is not easy.

- Biomass often has to be estimated – how would you measure the biomass of a whale?

- Usually a pyramid shows all the organisms at each **trophic level** of a **food web**, because animals rarely eat just one kind of food.

- It is usually impossible to measure all the organisms in a complete food web, so biomass is usually given in terms of an area, e.g. g/m^2.

- There is a time element too: think of the mass of food you have at home now, compared with the mass of food you will eat in a year, or your lifetime. How would the pyramid of biomass differ for each of these?

Each of these issues will affect the data that are collected.

Examiner feedback

Remember what you learned about the digestive system at Key Stage 3. It will help you understand this chapter, although you will not be examined on it at GCSE.

Natural recycling

A matter of life and death

The tissues of your body are made of many elements, which you get from your food. We call these elements nutrients because we need them for growth. A lack in any one of these over a long period would affect your growth and health.

Table 1 Important nutrients in a human body.

Nutrients	Percentage of body weight	Role in body
nitrogen	3.2%	forms part of proteins, DNA in all cells
calcium	1.8%	strengthens bones and teeth, needed for nerve and muscle activity
phosphorus	1.0%	part of DNA, also strengthens bones and teeth
sodium	0.2%	needed for nerve and muscle activity
iron	0.007%	haemoglobin in red blood cells carries oxygen

During photosynthesis plants make **carbohydrates**, which contain only carbon, hydrogen and oxygen. To grow well, the plant must have other nutrients, such as nitrogen for proteins and **DNA**, magnesium to make chlorophyll, and calcium to build strong cell walls. Plants absorb these nutrients when they take in water surrounding their roots. This removes the nutrients from the environment and makes it possible for animals to get what they need in their food. However, until the nutrients are returned to the soil, they are no longer available to plants.

Returning the nutrients to the soil only becomes possible when the plants or animals die, or when animals produce waste materials such as faeces and urine. The **decay** of this dead and waste material releases the nutrients back to the environment.

What is decay?

Decay is the digestion, or **rotting**, of complex **organic** substances to simpler ones by microorganisms such as fungi and bacteria. Since microorganisms have no gut, digestion happens outside their cells and the simpler nutrients are absorbed. This gives the microorganisms the nutrients they need for growth. Some of the simpler nutrients are left in the soil which means plants are now able to absorb them through their roots.

Fungus causing decay of an orange. The white part is a mass of fungal threads, and the grey/green parts are fruiting bodies.

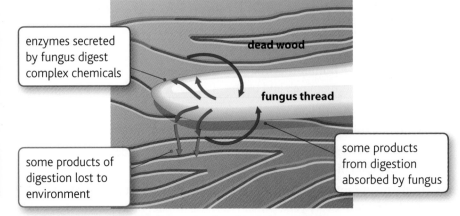

enzymes secreted by fungus digest complex chemicals

dead wood

fungus thread

some products of digestion lost to environment

some products from digestion absorbed by fungus

Figure 1 Decay organisms break down complex chemicals into smaller molecules that they can absorb.

The best conditions for decay

Decay microorganisms don't usually have a thick protective coating, so they easily dry out. They often grow below the surface of the material that they are decaying. However, many microorganisms are more active when there is plenty of oxygen for respiration, so they cannot live far beneath the surface. Like other organisms, the rate of the reactions in the cells of microorganisms is affected by temperature. When it is colder, they cannot grow as quickly, so decay is usually faster in warm, moist conditions.

a Which conditions would you test to investigate the best conditions for the decay of plant material?

b How would you carry out your investigation?

c How would you control variables?

In 1991, the body of a man was found in the Italian Alps after the deep ice above it had melted. The remains were dated at 3300 BC and still had skin and some hair.

Essential recycling

Without decay, there would be no recycling of nutrients in the environment, and there would eventually be no life. In **stable** natural communities, where there is little change over time in the plants and animals that live there, there is a balance between the nutrients removed from the soil by plants and the return of nutrients to the soil by decay. The nutrients are constantly cycled.

In winter, dead leaves cover the ground, but by midsummer most of the leaves will have decayed.

Questions

1 Describe one similarity and one difference in the way that microorganisms and animals digest their food.

2 Explain why the only parts of a fungal mould that you can usually see are the fruiting bodies.

3 The leaves in the forest photograph fell in November. Explain why they are still there in February but not in June.

4 **(a)** Draw a diagram to show the cycling of nutrients through the environment and living organisms.
(b) Use your diagram to explain the essential role of decay in nutrient cycling.

5 Put the following environments in order of decomposition rate of dead material, starting with the fastest, and justify your order: hot desert, Canadian coniferous forest, polar tundra, tropical rainforest.

6 Many gardeners put their garden waste on a compost heap to decay. Explain why the temperature inside the compost heap gradually increases over the first few weeks.

7 Explain fully why the 5000-year-old remains of the ice-man still had skin and hair.

8 Although rainforests are the most productive areas of plant growth on Earth, their soils are nutrient-poor. Explain as fully as you can the reasons for this apparent discrepancy.

Recycling issues

A load of rubbish

In 2009 a report on UK household waste estimated that we throw away about 8.3 million tonnes of food and drink each year. This is the equivalent of an average household throwing away £50 every month. Of the total, over 7.6 million tonnes goes in the waste-bin or down the drain.

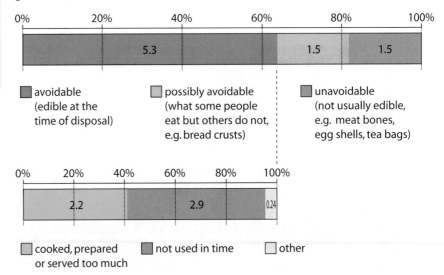

avoidable (edible at the time of disposal)

possibly avoidable (what some people eat but others do not, e.g. bread crusts)

unavoidable (not usually edible, e.g. meat bones, egg shells, tea bags)

cooked, prepared or served too much

not used in time

other

Figure 1 The amount of food and drink waste generated in the UK in 2009 (millions of tonnes per year).

The problem with organic waste

Food and drink wastes are organic, which means they originally come from plants and animals. These wastes decay as bacteria and fungi grow on them. Until recently much of the waste in the UK was disposed of in **landfill sites**. Once organic wastes are buried, the usual decay organisms cannot grow because there is not enough oxygen for aerobic respiration. Other decay microorganisms, called **methanogens**, can grow anaerobically. They release a gas called methane, which is not only highly flammable, but also more active in global warming than carbon dioxide.

In 1999, the European Landfill Directive set out plans to reduce the area used for landfill in all European countries for several reasons, including pollution. This has led to more waste recycling in the UK. Most councils now ask us to separate out garden waste, cans, glass and sometimes plastics for collection, but only a few collect food wastes separately. This is because meat waste is very attractive to pests like rats – the waste needs to be protected while it is waiting to be collected.

Landfill tips need to be vented in a controlled way for many years after tipping has ended, otherwise methane that forms as the refuse decays might explode or create fires that burn below the surface for weeks.

Collecting food waste separately means another separate bin for each house as well as special collection trucks.

A selection of choices

The traditional way of managing garden and vegetable kitchen waste is the **compost heap** in the garden. The waste is piled up and microbial decay breaks it down into compost that you can add to the garden soil. Many councils now collect this as 'green' waste and make compost on a large scale in a process called **windrow composting**. This needs a lot of space, and the composting material must be turned regularly to keep the oxygen level high, but it doesn't need any special equipment.

Other councils collect all garden and kitchen waste, including meat, together. By law, this needs composting in large containers until the meat waste is broken down, in a process called **in-vessel composting**. Composting is then completed using the windrow process. An advantage is that conditions can be monitored inside the containers, and maintained at the correct temperature and moisture levels for more rapid decay in the early stages. The higher temperatures also kill **pathogens** and the seeds of **weed** plants.

Another process uses **anaerobic digestion** by the methanogenic bacteria that cause problems on landfill sites. Food waste is put into large **digesters** and air is excluded. The methanogens break down the material and release methane and other gases. The methane is collected and burnt to produce heat, which can be used to heat buildings or for making electricity. This process can't use wood waste because these microorganisms cannot break it down.

In all these processes, the solid end materials can be used for soil conditioning, in gardens, parks or in agriculture.

Examiner feedback

You need to be able to compare different methods of dealing with organic waste, but the terms 'windrow composting' and 'in-vessel composting' do not need to be remembered for your exam.

A traditional compost heap in the garden can create the right conditions for vegetable waste to decay in a few months.

Questions

1   **(a)** What proportion of the food and drink that we waste in the UK could potentially be avoided? **(b)** Describe methods that would help us to reduce the amount of food and drink waste that we produce.

2   List the problems with disposing of food waste on landfill sites. Explain your choices.

3   Explain why meat waste must be composted in containers for the first part of in-vessel composting.

4   Explain how the recycling of food waste mimics the natural process of nutrient recycling.

5   Another way of dealing with all household waste is to burn it. Argue the environmental advantages of recycling food waste rather than burning it.

6   Draw up a table to show the different ways in which councils are managing waste food. For each way, list the advantages and disadvantages of each process.

7   Using the data in the graphs, create a poster to inform householders how and why they should increase the amount of waste they recycle to the maximum possible.

8   Some councils are considering fining households that put recyclable material into refuse that is not recycled. Prepare a memo for discussion listing the advantages and problems with this.

Science in action

Gardeners need to create the right conditions inside the compost heap to raise the temperature high enough to kill seeds of weed plants, and disease-causing fungi. So the heap must be kept sheltered from rain, and the right balance of fresh and dead material added plus a little soil to bring in the right microorganisms. Turning the heap every few weeks makes sure enough oxygen gets to all parts.

Route to A*

When evaluating a process, make sure you identify the advantages and disadvantages of the various approaches and then compare these to come to a decision about which gives the best result.

The carbon cycle

Learning objectives

- describe how plants remove carbon from the atmosphere and use it to make chemicals in their bodies
- explain how carbon passes between living organisms and is returned to the atmosphere
- interpret a diagram of the carbon cycle.

Figure 1 How carbon from the air is changed into carbon compounds in a potato plant.

carbon in carbon dioxide from respiration

carbon in carbon dioxide enters leaf and becomes part of glucose (photosynthesis)

carbon in protein and fats

carbon in starch and carbohydrates

Examiner feedback

Remember that plants photosynthesise during the day, taking carbon dioxide from the atmosphere, and plants respire all the time, releasing carbon dioxide back into the atmosphere.

Capturing carbon dioxide

Imagine a carbon atom that is part of a carbon dioxide molecule in the air. If that carbon dioxide molecule gets too close to a plant, it might be taken into a leaf. There, during photosynthesis, it will be changed into an organic carbon compound called glucose, a carbohydrate. This is known as **fixing** carbon because it removes carbon from the physical environment.

If the glucose is used for respiration, it will be converted to carbon dioxide and released back into the air. However, it might instead be changed into complex carbon compounds, such as the carbohydrates, proteins and fats in plant tissue.

Carbon compounds in animals

If a plant is eaten by an animal, the plant's tissues will be broken down during digestion. Some of the carbon compounds in it will be absorbed through the animal's gut and made into more carbohydrates, proteins, fats or other complex compounds. The rest will leave the animal's body as faeces.

Some of the carbon compounds made inside the animal will be converted to glucose for respiration, then transformed back into carbon dioxide, which will be released into the atmosphere. Other carbon compounds may become part of the animal's body tissues, or may be excreted in urine. If the animal is eaten by a predator, these processes will happen in that animal.

If plants and animals are not eaten, and just die, **detritus feeders**, such as worms and fly larvae, will feed on the dead bodies. They also feed on the faeces and urine excreted by animals. They break down the complex carbon compounds and use them to make more carbon compounds in their bodies, releasing some carbon as carbon dioxide to the air from respiration. Decomposer organisms, such as fungi and bacteria, continue the process of decay. They break down carbon compounds even further, using some to make carbon compounds in their bodies and using some for respiration, releasing carbon dioxide.

The full cycle

When large quantities of organisms, such as trees or plankton, are buried over long timescales, heat and pressure will change them into **fossil fuels**, such as coal and oil.

The **combustion** of fuels formed from organisms, including fossil fuels and wood, releases carbon dioxide into the air.

Whenever the carbon is returned to the atmosphere as carbon dioxide, it is possible that it may be captured by a leaf and fixed again in photosynthesis. This starts the process over again. This constant cycling of carbon through carbon compounds in living organisms and carbon dioxide in the air is called the **carbon cycle**.

In terms of chemicals and energy, we can see that although plants are able to take nutrients and carbon dioxide from the environment, and capture energy from sunlight, only the nutrients and carbon dioxide can be continuously cycled between the environment and living organisms. All the energy that is captured by plants will eventually be transferred as heat to the environment. This energy cannot be used by living organisms, so there is no energy cycle.

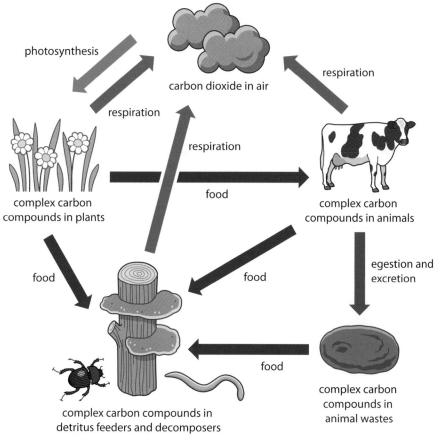

Carbon compounds in this cowpat will be broken down and used by many different kinds of detritus feeders and decomposers.

Figure 2 The natural carbon cycle.

Examiner feedback

Many different styles of carbon cycle diagram can be used in examination questions. Make sure that you have practised interpreting different styles of diagram.

Questions

1 What is photosynthesis, and what is its role in the carbon cycle?

2 What contribution do plants make to the carbon cycle?

3 Draw a diagram to show what happens to the carbon in an animal that is eaten by a predator.

4 It is possible that one of the carbon atoms in your body was once in the body of William Shakespeare. Explain how this could happen.

5 Explain how human activity modifies the carbon cycle shown in Figure 2.

6 In China, silkworms are grown on mulberry bushes that surround fish ponds. The faeces from feeding silkworms, and waste from the processing of silk, are returned to the ponds. Carp fish feed on water plants in the ponds, and are harvested for human consumption. Figure 3 shows the biomass in kilograms per hectare for the parts of this system. Describe what the diagram shows as fully as possible.

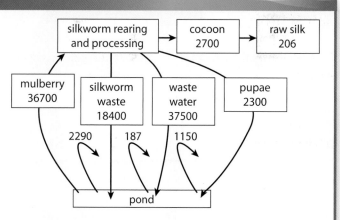

Figure 3 Biomasses.

7 The silkworm/carp pond arrangement is an example of a natural recycling system. Identify which parts of the system are not shown in the diagram and suggest as far as you can why they were not included.

8 Explain as fully as possible why the carbon cycling in the silkworm/carp pond arrangement is self-sustaining, and consider conditions in which this might change.

Assess yourself questions

1 The following methods are used to dispose of organic waste from the home and garden. Choose the correct method to answer each question. Each method can be used more than once.

   A   windrow composting
   B   anaerobic digestion
   C   landfill tipping
   D   in-vessel composting

   (a) Which method is *not* a form of recycling? *(1 mark)*

   (b) Which method is *not* suitable for meat wastes? *(1 mark)*

   (c) Which method produces methane and other gases that are used for making electricity? *(1 mark)*

   (d) Which method is *not* suitable for wood waste? *(1 mark)*

2 Dromedary (one-humped) camels survive better than any other large mammals in dry desert conditions.

   (a) Name two conditions of the environment that the camel must survive in a desert. *(2 marks)*

   (b) Explain how each of the adaptations shown in Figure 1 improves the camel's chances of surviving in the desert. *(4 marks)*

hump is a large store of fat

can drink much larger quantities of water at one go than most animals

body cells can tolerate higher levels of dehydration than most animals

wide flat feet

Figure 1 Camel adaptations.

3 **Table 1** The biomass of organisms at each trophic level living in a marsh.

Trophic level	Biomass / g/m²
producer	800
primary consumers	40
secondary consumers	10
tertiary consumers	2

(a) On graph paper, use these data to draw a pyramid of biomass. *(3 marks)*

(b) Define the terms *biomass* and *trophic level*. *(2 marks)*

(c) Explain why the data produce the shape of a pyramid. *(2 marks)*

(d) What is the source of the energy for the organisms in the marsh? *(1 mark)*

(e) What is the final form of the energy leaving the marsh? Explain your answer. *(2 marks)*

4 Figure 2 shows sewage from a pig farm discharging into a river. It also shows the samples of invertebrates that a scientist caught at different points along the river.

   (a) Which of these animals can only live in unpolluted water? *(1 mark)*

      (i)   rat-tailed maggot
      (ii) waterlouse
      (iii) mayfly nymph
      (iv) leech

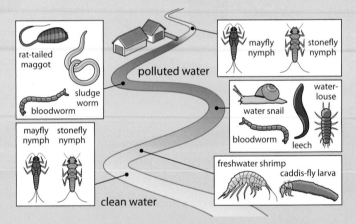

Figure 2 Map of river.

Figure 3 shows the concentration of oxygen in the water at different points along the river.

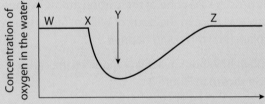

Figure 3 Oxygen concentrations.

(b) Which letter on the graph shows the position of the pig farm? *(1 mark)*

(c) Explain your choice in part b. *(2 marks)*

(d) The bloodworm is a water pollution indicator. Explain what this means. *(1 mark)*

(e) Describe one adaptation of the bloodworm to living in polluted water. *(1 mark)*

5   The herring gull and the Arctic skua are birds that
    breed in the UK. Figure 4 shows the change in breeding
    population size for these species since 1986.

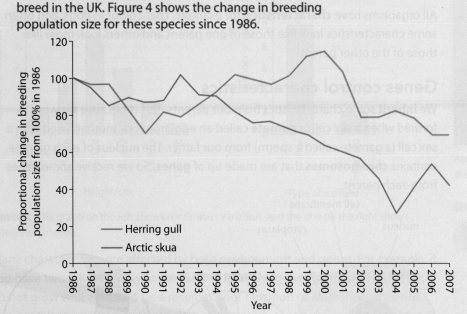

Figure 4 Herring gull and Arctic skua poulation.

Arctic skuas breed only in the north of Scotland and
have a population size of about 2100 pairs. Herring
gulls breed in many parts of the UK and there are an
estimated 130 000 pairs. Both species were put on the
'red' list (birds of concern) in 2009 because of their
decline in population size.

Arctic skuas depend on sand eels to feed their chicks.
In recent years, the numbers of sand eels have greatly
decreased. Some scientists think this is related to an
increase in temperature of the North Sea over the past
decade. Herring gulls feed their chicks a wide range of
food, including fish, worms from fields and scraps from
landfill tips.

(a) How does the population change for each of the
     species between 1986 and 2007?   *(2 marks)*

(b) Some people say that climate change is affecting
     the distribution of animals.

   (i) Is this a reasonable explanation for the decline
        in skuas? Explain your answer.   *(1 mark)*

   (ii) Is this a reasonable explanation for the decline
        in herring gulls? Explain your answer.   *(1 mark)*

(c) Since 2002, there has been an increase in the
     amount of food and plant waste that is treated at
     recycling centres.

   (i) Explain how this could affect herring gulls.
                                          (1 mark)

   (ii) Does the graph support this argument? Explain
        your answer.   *(1 mark)*

(d) Do you think that the decline of each of these
     species should be 'of concern'? Explain your answer.
                                          (4 marks)

6   Compare the contributions of plants, animals and
    decomposers to the carbon cycle.

    *In this question you will be assessed on using good English,
    organising information clearly and using specialist terms
    where appropriate.*   *(6 marks)*

7   During a survey for banded snails, students each
    searched a different habitat until they had found
    15 snails.

Table 2 The students' results.

Snail colour and banding	Habitat		
	long grass	woodland	short grass
brown with bands	1	1	3
yellow with bands	9	2	4
brown no bands	1	9	1
yellow no bands	4	3	7

Assuming that predation was the main factor
controlling distribution, the students had predicted that
more banded snails would be found in long grass than
in the other habitats, where the bands would give better
camouflage in stripy shadows. They also predicted that
more brown snails would be found in deep shade.

(a) Evaluate the method and suggest how it could be
     improved.   *(2 marks)*

(b) Analyse the results and explain whether they
     support the two predictions.   *(2 marks)*

(c) What other evidence should the students look for
     to support their idea that predation is a major factor
     controlling distribution in these areas? Explain your
     answer.   *(2 marks)*

Different types of reproduction

Learning objectives

- describe the key features of sexual and asexual reproduction
- explain how sexual reproduction leads to variety in the offspring and why asexual reproduction produces identical offspring
- explain the advantages and disadvantages of the different forms of reproduction.

Sexual reproduction

The variation described in lesson B1 6.1 is the result of **sexual reproduction**. This kind of reproduction occurs when the nucleus of a sperm cell (male gamete) fuses with the nucleus of an egg cell (female gamete) to produce a fertilised cell. When the gametes are formed, they only receive half of the chromosomes that are in a body cell. So the fertilised cell gets half its chromosomes from the mother and half from the father. The process of producing gametes ensures that it is highly unlikely that any two gametes will contain the same variations of genes.

Parent cell containing pairs of chromosomes; the two chromosomes in a pair contain different versions of the same genes.

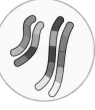

(*Note*: this diagram is highly simplified.)

Four possible different sex cells.

Figure 1 Each sex cell gets one chromosome from each pair in the parent cell.

The variation in the genes in each gamete, and the mixing of genes from the father and mother, make it very unlikely that two offspring from the same parents will be alike, unless they are identical twins formed from the same fertilised cell.

Asexual reproduction

Another way to produce offspring is by **asexual reproduction**. Here the offspring are produced from the division of cells in the parent without the need for **fertilisation** by a sperm cell. So the cells of the offspring contain exactly the same chromosomes as the parent. All the offspring are genetically identical and are called **clones**.

Asexual reproduction can happen much more quickly than sexual reproduction. For example, in one summer, there may be up to 40 generations of cabbage aphid. This is a great advantage for a pest such as aphids that need to reproduce rapidly in the short time when food is available.

Some plants also reproduce asexually. For example, couch grass produces new plants from underground stems, called **runners**, which help it spread quickly in an area to outcompete other plants. Plants that live in places where it is either too hot or too cold for growth during part of the year, may form underground **storage organs**, like potatoes, each of which can form a new plant in the next growing season.

Many species of plants and animals, like these damselflies, use sexual reproduction to produce offspring.

Female aphids can give birth to live young without fertilisation by a male aphid.

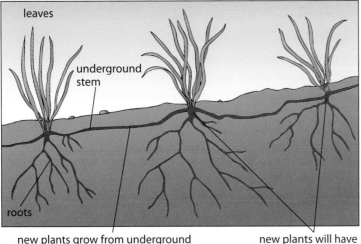

leaves

underground stem

roots

new plants grow from underground stems by asexual reproduction

new plants will have identical genes

Figure 2 When the new couch grass plant has enough roots to grow on its own, the runner may die off.

A tray of leaf cuttings from an African violet plant. The dark leaves were taken from the parent plant. Some of these have formed roots and are starting to grow into new plants with brighter green leaves.

Applications of asexual reproduction

We can make clones of plants artificially by copying the natural asexual process, such as taking potatoes from an old plant and planting them separately. For some plants we can also take **cuttings**. These are parts cut off a plant, usually from a stem or leaf. The cuttings grow roots and develop into new plants. This can produce lots of new plants more quickly than by sexual reproduction, and more cheaply because it is quicker.

Questions

1 Explain why the offspring from sexual reproduction are not identical to: **(a)** their parents **(b)** other offspring from the same parents.

2 Explain why the offspring from asexual reproduction are clones.

3 Draw up a table to show the advantages and disadvantages of sexual and asexual reproduction.

4 A student grows African violet plants. Which technique should he use to: **(a)** try to produce a plant that has a new flower colour? **(b)** produce many plants with this same new colour? Explain your answers.

5 Aphids reproduce asexually all summer, but late in the season winged males and females are produced. These mate and the females lay eggs that overwinter and hatch in spring. **(a)** Are females that hatch from overwintering eggs identical or not? Explain your answer. **(b)** What is the advantage of producing aphids that are able to fly in autumn? **(c)** Explain the advantages of both kinds of reproduction in the aphid life cycle.

6 Use your table from question 3 to suggest the best form of reproduction in the following situations. Justify your answers. **(a)** A weed plant starts growing in cleared soil. **(b)** An aquatic animal produces offspring that will drift downriver to live in other areas.

7 A student has been given a *Bryophyllum* plant to set up an experiment to show as clearly as possible the variation in leaf size caused by temperature. Write a plan for how she should set up the experiment, explaining each step.

A*

Cloning plants and animals

Tissue culture

Plant growers have grown new plants from cuttings for many centuries. Now they can also grow new plants using just a few cells from the parent plant. This is known as **tissue culture** and is another form of cloning.

Cells are taken from the tip of a shoot and placed on a jelly that contains nutrients and a chemical that helps the cells to divide. They make a small ball of cells called a **callus**. The callus can be split to make new calluses. Each callus is then put on a jelly that contains different chemicals to encourage roots and shoots to form. When the new plants are large enough, they are planted into compost. Tissue culture makes it easier to grow thousands of new plants from one original one.

Embryo transplants

When animals reproduce, most of the cells in the embryo **specialise** before the animal is born. Specialised animal cells, such as muscle cells, cannot change into other kinds of cell. This means that cloning animals is much more difficult than cloning plants, but it can be done.

One way is called **embryo transplanting**. An egg is fertilised with sperm in a laboratory. When it has divided to make four or eight cells, before they start to specialise, the cells are separated to start making new embryos. These are transplanted into the womb of **host mothers** where they grow until they are ready to be born.

Forming new plants in tissue culture.

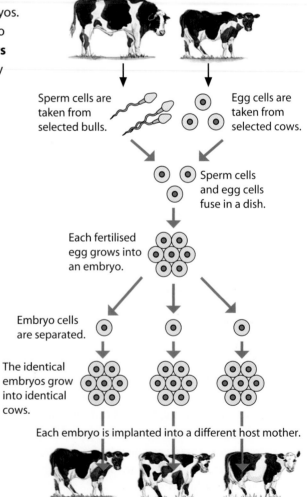

Sperm cells are taken from selected bulls.

Egg cells are taken from selected cows.

Sperm cells and egg cells fuse in a dish.

Each fertilised egg grows into an embryo.

Embryo cells are separated.

The identical embryos grow into identical cows.

Each embryo is implanted into a different host mother.

Figure 1 Embryo transplanting means farmers can get many more offspring from their best animals.

Adult cell cloning

Another animal cloning technique is **adult cell cloning**, where the nucleus of an unfertilised egg cell is removed and replaced with the nucleus of a body cell, e.g. skin cell, from an adult animal. The egg cell can then be given an electric shock so that it starts to divide like a normal embryo. The embryo will contain the same genetic information as the adult body cell. Although this was first done successfully with sheep, it has been repeated for many animal species. People are concerned that it could be used to make human clones, but scientists are using the technique to make replacement cells for problems such as spinal cord damage.

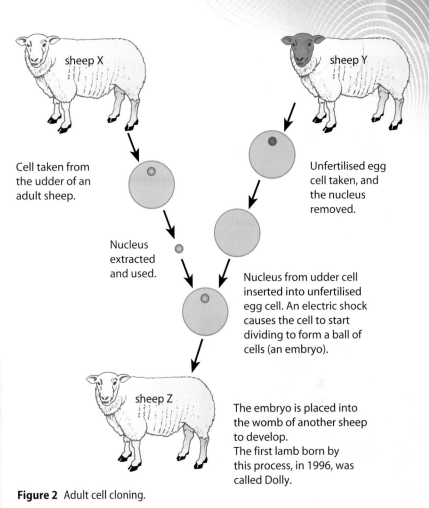

Cell taken from the udder of an adult sheep.

Nucleus extracted and used.

Unfertilised egg cell taken, and the nucleus removed.

Nucleus from udder cell inserted into unfertilised egg cell. An electric shock causes the cell to start dividing to form a ball of cells (an embryo).

The embryo is placed into the womb of another sheep to develop.
The first lamb born by this process, in 1996, was called Dolly.

Figure 2 Adult cell cloning.

Science skills

It took 277 attempts using adult cell cloning to create a cell that successfully grew into a healthy sheep. Most attempts fail because the cells do not develop normally. As an adult, Dolly was mated and produced healthy lambs, but she developed arthritis (usually occurs in much older sheep) and was put down at the age of six because of lung cancer. The scientists who bred her said her early death was not the result of being a clone, but other scientists think that using an 'old' nucleus could cause the animal to age faster than normally.

a What are the pros and cons of continuing adult cell cloning research?

Science in action

Since Dolly the sheep was created, the same technique has been used in many other species, including cats and dogs. For about US$50 000 you could have your pet dog or cat cloned. However, although the clone would look like your pet, it probably wouldn't behave exactly the same way.

Questions

1 **(a)** How is tissue culture different from taking cuttings? **(b)** How is tissue culture the same as taking cuttings?

2 Are all animals produced by embryo transplanting clones? Explain your answer.

3 Why would it make sense for a farmer to use an expensive technique such as embryo transplanting rather than allowing bulls and cows to mate as usual?

4 Look at the diagram in Figure 2, which shows how Dolly the sheep was produced. The three adult sheep were all different species. Which sheep was Dolly a clone of? Explain your answer.

5 Draw a flow chart to show all the steps in tissue culture.

6 Describe how adult cell cloning could be used to create a human clone.

7 Tissue culture is being used increasingly to save rare and endangered plant species. Explain as fully as you can the advantages and disadvantages of using tissue culture for this, rather than collecting seed as scientists used to do.

8 Evaluate the disadvantages of producing a human clone using adult cell cloning.

Modifying the genetic code

Genetic engineering

The genetic information in all organisms works in the same way, whether it comes from a plant, an animal or a bacterium. Therefore, we can take a gene for a particular characteristic from a chromosome of an individual of one species and insert it into an individual of a different species and it will produce the same characteristic. This is called **genetic engineering**. The organism that contains the new gene has been **genetically modified** (GM), and is called a **transgenic organism**.

Genetic engineering has many practical applications. For example, we can transfer the gene for making the hormone insulin from a human cell into a bacterium. Insulin is a hormone that is needed by some diabetic people to prevent them becoming very ill. It is possible to grow many genetically modified bacteria on a large scale in bacterial fermenters. This means we can make a lot of human insulin more cheaply and safely than before. We now use GM bacteria to make human growth hormone and a vaccine to protect against infection by the disease hepatitis B.

This mouse contains a gene from a fluorescent jellyfish that glows under blue light.

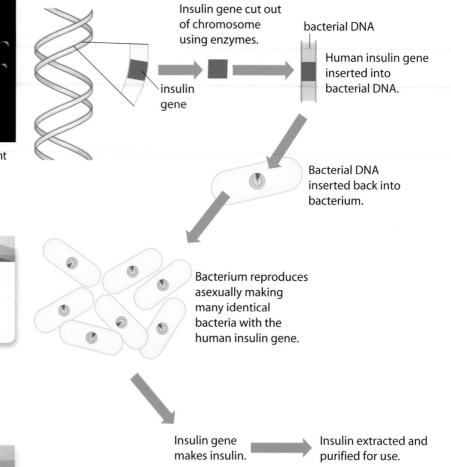

Insulin gene cut out of chromosome using enzymes.

insulin gene

bacterial DNA

Human insulin gene inserted into bacterial DNA.

Bacterial DNA inserted back into bacterium.

Bacterium reproduces asexually making many identical bacteria with the human insulin gene.

Insulin gene makes insulin.

Insulin extracted and purified for use.

Figure 1 How genetically modified human insulin is produced.

Before we could make GM bacteria to produce human insulin, the insulin was extracted from dead animals, such as pigs. This not only meant there was less insulin available, but the insulin produced was not identical to human insulin and could cause health problems.

It is also possible to insert a gene into some of the body cells rather than an early embryo. This means that only those body cells make what the gene codes for, and the gene cannot be passed on to offspring. Scientists are developing ways of treating human diseases caused by faulty genes, such as cystic fibrosis, in this way.

Transferring genes at an early stage

Genes can also be transferred into a plant or animal embryo at an early stage in their development. As the organism grows, all the cells in its body will have a copy of the inserted gene, so all cells can develop the desired characteristics. This technique is most commonly used to make GM food plants. However, it is also very important in research for causes and treatments of human diseases. Mice have been genetically modified so that they can be used to study human cancers, or to investigate the effect of changes to particular genes that cause human diseases. Using mice like this has rapidly increased our knowledge and means treatments will be developed sooner.

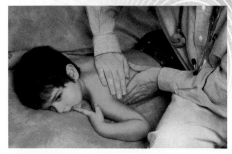

People who have cystic fibrosis need treatment every day to stop them becoming very ill. In the future they might be able to insert the correct genes into their lung cells using an inhaler.

This mouse is genetically modified so that it can be used to find out more about human cancers, how they are caused and how they can be treated.

Science skills

Inserting genes into human embryos is illegal. Although it could help cure genetic diseases, there are many arguments against it:

- treatment would be expensive, so would only wealthy people get it?
- the parents choose the treatment not the child who gets it – is this right?
- what if parents could choose other genes for insertion, such as for height or intelligence?

Questions

1   Define the term *genetic engineering* in your own words.

2   Explain why we can transfer a characteristic from one species into a different species.

3   Suggest why the insertion of genes into a human embryo is illegal in most countries.

4   Draw a flow chart to describe how human growth hormone could be produced by genetically modifying bacteria.

5   List as many advantages as you can for producing human insulin from genetically engineered bacteria.

6   Should a parent have the right to choose to have a gene inserted into their embryo so that their child does not have a genetic disease? Justify your answer.

7   Compare as fully as you can the effect of inserting genes into body cells with the effect of inserting genes into embryo cells.

8   Evaluate the ethical and social issues of using GM mice to research the causes and treatment of human diseases such as cancer.

Examiner feedback

It is important that you appreciate the ethical concerns some people have about gene therapy and why it continues to be the focus of much debate.

Taking it further

Plasmids from bacteria are one of the main methods scientists use for transferring genes into bacteria to make genetically modified bacteria. This process copies the natural process that many types of bacteria use for exchanging genetic material.

Making choices about GM crops

Making GM crops

Plants can be genetically modified by inserting the required gene into the cells of an early embryo. Often a bacterium, called *Agrobacterium*, is used to get the new gene inside the nucleus of the cells where it can join with the cell's DNA.

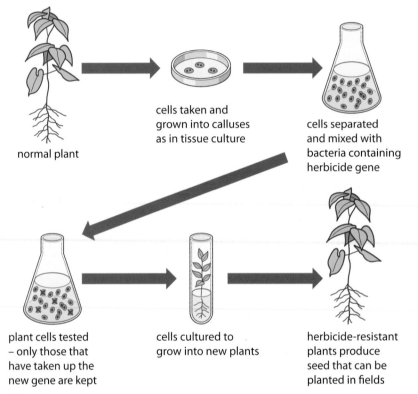

normal plant

cells taken and grown into calluses as in tissue culture

cells separated and mixed with bacteria containing herbicide gene

plant cells tested – only those that have taken up the new gene are kept

cells cultured to grow into new plants

herbicide-resistant plants produce seed that can be planted in fields

Figure 1 Producing GM plants.

Different modifications

Many GM crops have a gene for **herbicide** resistance. While a crop is growing, other plants compete for water and nutrients in the soil. These weed plants can reduce the amount of food harvested from the crop (the **yield**). Farmers use herbicides to kill plants, but these chemicals can damage the crop plants as well. Using crops that are resistant to a particular herbicide means that the crop can be sprayed with that herbicide to kill weeds without damaging the crop.

Some crop plants are modified with a gene for a poison that kills insects that try to eat the plant. Pest damage reduces crop yield, so GM crops should produce a greater yield.

The pros and cons of GM crops

One concern with GM crops is that the seed costs more than normal seed, and the companies that produce seed for herbicide-resistant crops also produce the herbicide that the crop is resistant to. This is good for those companies but not necessarily good for farmers.

In Africa, the maize stalk borer damages plants and reduces crop yields by 20–40% on average.

Another concern is gene transfer through **pollination**. Pollination can happen occasionally between plants that are closely related. As crop varieties were bred originally from wild varieties, if those wild types are weeds growing nearby, then the inserted gene could be transferred to them. Research in Canada has shown that the gene for herbicide resistance has been transferred to weed species within six years of growing GM crops.

Using GM varieties changes the way farmers look after their crops, which can also affect the environment. Growing insect-resistant crops can benefit other species of insects because chemical insecticides, which kill other insect species as well as the pest, are not used. However, where herbicide-resistant varieties have been grown on a large scale, there is no evidence of reduced use of chemicals that kill a wide range of plant species and this means that the variety of wildflowers is reduced, which will affect insect and bird species that feed on them.

People are also concerned about the safety of the foods produced from the crops. Although the foods are tested on animals to make sure they don't cause health problems, we have no idea yet if there are any problems caused by eating GM foods over a long time.

Much more research is needed if GM crops are to help feed the increasing human population.

Spraying a field with herbicide will get rid of not just the weeds, but also the animals that feed on the weeds, and animals that feed on those animals.

Opinion on the value of GM crops is divided because people are not sure about long-term effects.

Questions

1 Explain why plant embryos are used for genetic modification.

2 Sketch a diagram to show how weeds can reduce crop yield.

3 Suggest any disadvantages of using crops that are modified to resist insect attack.

4 Explain how using an insect-resistant crop variety could help the environment.

5 No GM crops are grown for sale in the UK at the moment, but there are licensed scientific trials. Should this be allowed? Justify your answer.

6 A small-scale farmer in Kenya is considering growing GM maize that is resistant to stalk borer. How would you evaluate the advantages and disadvantages of this so that you could advise on the best choice?

7 Genetic engineering and tissue culture are two possible solutions to the problem of creating disease-free bananas. Explain fully why using tissue culture is more likely to be the better approach for poor farmers in Africa.

8 Explain as fully as you can why results for yield in trials on GM crops by the companies that make the seed might be more positive about increased yields than the results from other scientists.

Examiner feedback

You will be expected to understand and present both sides of the ethical debate on GM crops, and not just give your own point of view.

You will be expected to apply the knowledge and understanding you have gained on your course to new examples.

ISA practice: the growth of mould on bread

Scientists are investigating the best advice to give to supermarkets about storing and displaying bread. Your task is to investigate the effect of temperature on the growth of mould on bread.

Hypothesis

It is suggested that there is a link between temperature and the rate at which bread goes mouldy.

Section 1

1   In this investigation you will need to control some of the variables.

(a) Name one variable you will need to control in this investigation. *(1 mark)*

(b) Describe briefly how you would carry out a preliminary investigation to find a suitable value to use for this variable. Explain how the results will help you decide on the best value for this variable. *(2 marks)*

2   Describe how you are going to do your investigation. You should include:

- the equipment that you would use
- how you would use the equipment
- the measurements that you would make
- how you would make it a fair test.

You may include a labelled diagram to help you to explain your method.

In this question you will be assessed on using good English, organising information clearly and using specialist terms where appropriate. *(6 marks)*

3   Think about the possible hazards in your investigation.

(a) Describe one hazard that you think may be present in your investigation. *(1 mark)*

(b) Identify the risk associated with the hazard you have described, and say what control measures you could use to reduce the risk. *(2 marks)*

4   Design a table that will contain all the data that you are going to record during your investigation. *(2 marks)*

Total for Section 1: 14 marks

Section 2

A group of students, Study Group 1, investigated the effect of temperature on the growth of mould on bread.

- Their teacher gave them two slices of bread, each of which had a colony of mould growing on it.
- They measured the width of the colonies then placed each of the slices of bread inside separate plastic bags. They placed one plastic bag a shelf in the laboratory. They placed the other plastic bag in a refrigerator.
- On each of the next four days, they removed the slices of bread from the bags, measured the width of the mould colonies then replaced the bread in the bags. The bags were returned to the shelf in the laboratory and the refrigerator respectively.

Figure 1 shows their results.

bread kept in laboratory		bread kept in refrigerator
4 cm	Day 1	4 cm
6 cm	Day 2	5 cm
10 cm	Day 3	6 cm
15 cm	Day 4	7 cm
20 cm	Day 5	8 cm

Figure 1 Group 1's investigation.

5   (a) (i) What is the independent variable in this investigation?

(ii) What is the dependent variable in this investigation?

(iii) Name one control variable in this investigation. *(3 marks)*

(b) Plot a graph to show the link between temperature, time and the diameter of the mould colony. *(4 marks)*

(c) Do the results support the hypothesis? Explain your answer. *(3 marks)*

Below are the results of three other study groups.

Table 1 shows the results of another group of students, Study Group 2.

Table 1 Results from Study Group 2.

Temperature at which bread was stored/°C	Number of days the bread stayed mould-free
0	10.0
10	5.0
20	3.5
30	2.4
40	1.6
50	1.0

A third group of students, Study Group 3, also investigated the hypothesis. Figure 2 is a graph of their results.

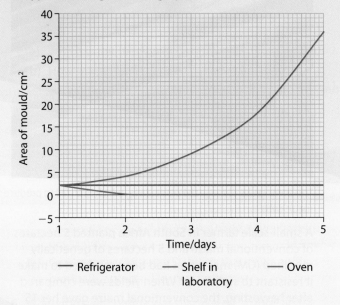

— Refrigerator   — Shelf in laboratory   — Oven

Figure 2 Study Group 3's results.

Study Group 4 was a group of scientists investigating the best conditions for storing bread in supermarkets. They investigated the effect of temperature on the rate at which the bread went stale. Figures 3 and 4 show their results.

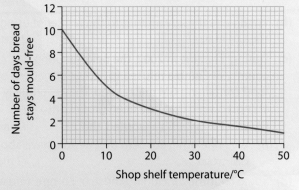

Figure 3 Study Group 4's results: effect of temperature on the rate at which bread goes mouldy.

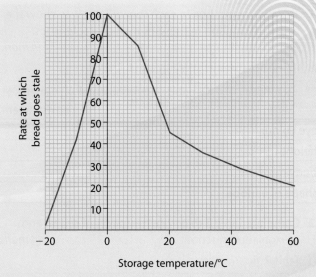

Figure 4 Study Group 4's results: effect of storage temperature on rate at which bread goes stale.

6 Describe one way in which the results of Study Group 2 are similar to or different from the results of Study Group 1, and give one reason why the results are similar or different. *(3 marks)*

7 **(a)** Draw a sketch graph of the results from Study Group 2. *(3 marks)*

   (b) Does the data support the hypothesis being investigated? To gain full marks you should use all of the relevant data from the first set of results and Study Groups 2 and 3 to explain whether or not the data supports the hypothesis. *(3 marks)*

   (c) The data from the other groups only gives a limited amount of information. What other information or data would you need in order to be more certain as to whether or not the hypothesis is correct? Explain the reason for your answer. *(3 marks)*

   (d) Use the results from Study Groups 2, 3 and 4 to answer this question. What is the relationship between the temperature and the growth of mould on bread? How well does the data support your answer? *(3 marks)*

8 Look back at the investigation method of Study Group 1. If you could repeat the investigation, suggest one change that you would make to the method, and give a reason for the change. *(3 marks)*

9 Suggest how ideas from your investigation and the scientists' investigations could be used to advise supermarkets about the best ways of storing and displaying bread. *(3 marks)*

Total for Section 2: 31 marks
Total for the ISA: 45 marks

Here are three students' answers to the following question:

During its life the rabbit ate a lot of grass. The carbon in the grass became part of the rabbit's body. The rabbit has died and is left and not eaten. How is the carbon recycled back into the grass which other rabbits will eat?

In this question you will be assessed on using good English, organising information clearly and using specialist terms where appropriate. (6 marks)

Read the three different answers together with the examiner comments. Then check what you have learnt and put it into practice in any further questions you answer.

B Grade answer

Student 1

> The correct scientific word is decay.

> It is better to say that the carbon dioxide is released.

Grass photosynthesises which means it takes in carbon dioxide from the air. The rabbit ate the grass which contained carbon and the rabbit used this to make carbohydrates. The rabbit is now dead so that means its body will start to rot. Microorganisms are involved in this process. The carbon in the grass that the rabbit ate will go back into air as carbon dioxide.

Other rabbits will come along and eat the grass and the whole process will start again.

Examiner comment

This candidate has understood that the process is a cycle, but they should have started with the dead rabbit and ended up with photosynthesis. They have not explained how the rabbit decays, including the role of microorganisms. Microorganisms respire and enzymes are involved; grass photosynthesises and takes in carbon dioxide to make carbohydrates for it to grow; the rabbit eats the grass and takes in carbon; so as the rabbit decomposes carbon dioxide is released into the atmosphere.

A Grade answer

Student 2

> To synthesise carbohydrates would be better.

> It would have been better to use the more scientific terminology digest.

The rabbit ate and digested the grass which contained carbon. The grass photosynthesised which means it used carbon dioxide from the air to allow it to grow. Now that the rabbit is dead its body will start to decay because microorganisms will eat it. The microorganisms respire so they will use the carbohydrates they have eaten to produce energy. This process releases carbon dioxide back into air.

Examiner comment

Like student 1, this candidate has understood the carbon cycle. However, they have also discussed how the rabbit decays and that microorganisms are involved. They have also pointed out that microorganisms respire, releasing carbon dioxide back into the atmosphere. They should have started with the rabbit decaying first.

A* Grade answer

Student 3

> Use of correct terminology.

> Respiration of microorganisms correctly mentioned.

When a rabbit dies its body is decomposed by soil microorganisms. Some of these microorganisms use enzymes to digest the complex carbon compounds in the rabbit's body into sugars. These sugars are then used by the microorganisms in respiration to produce energy. The carbon dioxide produced during respiration is released into the atmosphere.

Grass absorbs carbon dioxide from the atmosphere during photosynthesis, producing carbohydrates, which contain carbon. When rabbits eat grass, the carbohydrate in the grass is digested, forming sugars. The sugars are absorbed into the rabbit's blood.

> carbohydrates produced by photosynthesis is correct.

Examiner comment

This candidate has covered all the main points in the process: microorganisms use the energy from respiration to decay/digest the rabbit. As they do this, they respire and release carbon dioxide into the atmosphere. Enzymes are involved in this process. Grass uses the carbon dioxide from the atmosphere in photosynthesis to convert the carbon into carbohydrates to help it grow. Rabbits eat the grass, and digest the carbohydrates in the grass.

MOVING UP THE GRADES

- Read the question carefully.
- These questions carry a maximum of either five or six marks.
- Plan your answer by noting at least five/six relevant points you are going to make.
- Put these points into a logical sequence.

4   The diagram shows a food chain from the Antarctic ocean.

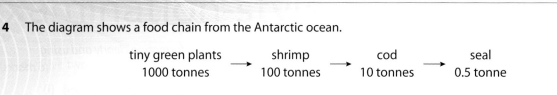

tiny green plants → shrimp → cod → seal
1000 tonnes   100 tonnes   10 tonnes   0.5 tonne

The cod is a fish and the seal is a mammal.

(a) Draw and label a pyramid of biomass for this chain.   *(2 marks)*

(b) The ratio of the biomass of shrimp to the biomass of the cod is much less than the ratio of the biomass of the cod to the biomass of the seal. Explain why.   *(3 marks)*

5   The diagram shows the mass of carbon involved each year in some of the processes in the carbon cycle.

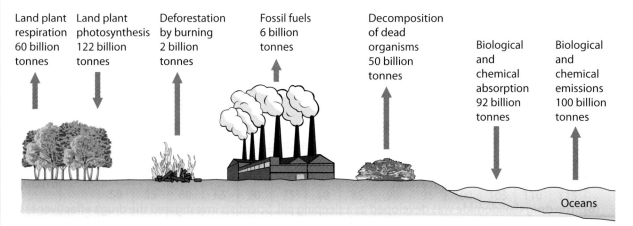

Land plant respiration 60 billion tonnes

Land plant photosynthesis 122 billion tonnes

Deforestation by burning 2 billion tonnes

Fossil fuels 6 billion tonnes

Decomposition of dead organisms 50 billion tonnes

Biological and chemical absorption 92 billion tonnes

Biological and chemical emissions 100 billion tonnes

Oceans

(a) Calculate the net change to the mass of carbon in the atmosphere in one year.   *(2 marks)*

(b) Decomposition of dead organisms releases 50 million tonnes of carbon into the atmosphere every year.
Explain how decomposition of dead organisms releases carbon into the atmosphere.   *(2 marks)*

6   Lichens are indicators of the concentration of sulfur dioxide in the atmosphere.

Students investigated the distribution of lichens in and around a city. From their results the students divided the city and its surroundings into six zones.

Table 1 shows the range of sulfur dioxide concentrations in the six zones.

Table 2 shows which species of lichen are found in each zone.

Table 1

Pollution zone	Sulfur dioxide concentration / $\mu g/m^3$
1	greater than 75
2	55 to 75
3	45 to 54
4	35 to 44
5	10 to 34
6	less than 10

Table 2

Species of lichens	Pollution zone
Desmococcus viridis	1
Evernia prunastri	2
Physcia adscendens	2
Xanthoria polycarpa	2
Physcia aipolia	3
Melanelia species	3
Cetraria chlorophylla	4
Graphis scripta	4
Menegazzia terebrata	5
Parmotrema arnoldii	5
Usnea rigida	5
Lobaria pulmonaria	6

(a) Which lichen species can tolerate the highest concentration of sulfur dioxide? *(1 mark)*

(b) What is the maximum sulfur dioxide concentration that *Graphis scripta* can tolerate? *(1 mark)*

(c) To carry out their survey, the students travelled by bus. They got off at each bus stop, measured the sulfur dioxide concentration with a meter and identified the lichens growing within 10 metres of the bus stop.

Evaluate the method used by the students to collect their data. *(3 marks)*

(d) Which is the better indicator of sulfur dioxide pollution in an area, using a meter or doing a lichen survey?

Explain the reason for your answer. *(2 marks)*

7   The vole is a small mammal, about the size of a house mouse.

Scientists investigated the size of voles in different parts of Europe. They found that voles in Northern European countries were larger than those found in southern Europe.

Suggest an explanation for the evolution of the differences in size of the voles.

In this question you will be assessed on using good English, organising information clearly and using specialist terms where appropriate. *(6 marks)*

8   Scientists have discovered genes that make plants resistant to attack by fungi in close relatives of potato plants. They have successfully transferred these genes into potato plants. The scientists are planning to grow a trial crop of the genetically engineered potatoes.

(a) (i) Describe how the scientists remove genes from a relative to a potato plant. *(2 marks)*

(ii) Describe how scientists might clone the genetically engineered potato plant. *(2 marks)*

(b) Evaluate the issues surrounding the growing of a pilot crop of genetically engineered potato plants.

In this question you will be assessed on using good English, organising information clearly and using specialist terms where appropriate. *(6 marks)*

Growing and using our food

Cells of animals, plants and bacteria have a characteristic structure, typical of their type. Multicellular organisms, like humans, are made up of cells that have differentiated to perform different functions. The digestive system, for example, includes many specialised cells grouped into different types of tissues. Organs like the stomach and small intestine are each made up of several different tissues.

Green plants use light energy to make their own food during photosynthesis. This is the basis of our agriculture. Some of our food crops are grown in greenhouses or polytunnels where the environment can be controlled. Light, temperature and carbon dioxide concentration can be manipulated to enhance the crop's growth.

In natural environments, physical factors, such as amount of light and availability of water, affect where we find certain plants. Fieldwork using quadrats and transects can provide quantitative data about the types and number of plants growing in a habitat. A tree trunk's surface, the ground in the shade of a tree, or a field can be investigated.

Test yourself

1. Where are chromosomes found in plant and animal cells?
2. Describe the reproduction rate of bacteria and explain whether bacteria reproduce faster in the human body or at a room temperature of 20 °C.
3. Explain what is meant by a healthy diet.
4. Why are organisms able to survive best in conditions in which they normally live?

Objectives

By the end of this unit you should be able to:

- describe the structure and function of animal, plant, bacterial and yeast cells
- explain what is meant by diffusion and how it is affected by the difference in concentration between two areas
- list the hierarchy of specialised cell organisation into organ systems, such as the digestive system
- explain that the salivary glands, stomach, pancreas and small intestine contain glandular tissue that produces digestive juices
- describe how the limiting factors of photosynthesis can be controlled to enhance crop growth in artificial environments
- tabulate the uses of glucose in plants and algae
- interpret the distribution of organisms in a community by referring to the physical factors that may affect them.

Animal building blocks

Learning objectives

- describe the structure and function of the parts of animal cells
- explain what controls chemical reactions in cells
- explain how animal cells may be specialised to carry out a particular function.

Animal cells

Animals, including humans, are made up of millions of tiny cells. You can see some of the structure of these cells through a light microscope that can magnify up to 400 times.

Each cell is surrounded by a very thin **cell membrane** that holds the cell together. The cell membrane also controls what goes into and out of the cell.

Cells contain smaller parts called **organelles**. These include the **nucleus**, **mitochondria** and **ribosomes**. The organelles have particular jobs in the cell. The single nucleus controls the cell's activities and it is surrounded by watery **cytoplasm**. Without the nucleus the cell will die. It contains **DNA**, the **genetic** material that provides the instructions for synthesising the chemicals the cell needs, like **enzymes**. Inside the cytoplasm hundreds of chemical reactions take place and these reactions are controlled by the enzymes.

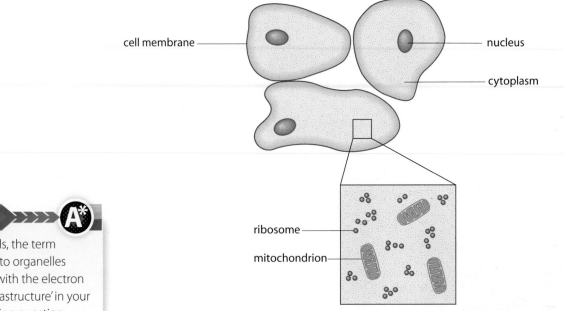

Figure 1 The organelles found in animal cells.

Electron microscopes magnify up to 500 000 times and show us details of smaller organelles, such as mitochondria. These use **glucose** in respiration to release energy for the cell. Ribosomes are the smallest organelles. They build up, or synthesise, **proteins** from smaller, simpler compounds called **amino acids**. Proteins are molecules that are used to make other parts of the cell and other chemicals, such as enzymes. The cells of most animals, including humans, have the same organelles.

Specialised animal cells

The cells in Figure 1 line the inside of your mouth. They are called simple **epithelial cells** and line cavities and tubes, like blood vessels in your body. Many different types of cell are found in your body. They have different shapes and many have special features that are related to what they do. These cells are called **specialised cells**. Some examples are shown in Figure 2.

a Spindle-shaped muscle cells have fibrils and can shorten in length.

fibrils

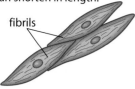

b Sperm cells have a tail to help them move to find the egg. They also have a high number of mitochondria to release energy for movement.

tail

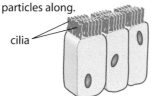

c Nerve cells have long fibres that carry electrical impulses. Branches of cytoplasm at each end of the cell facilitate communication with other nerve cells.

nerve fibre

cytoplasmic branches

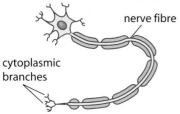

d These specialised epithelial cells have tiny hair-like structures, called cilia, on their free surface. They are known as ciliated epithelial cells. The cilia sway constantly back and forth to move particles along.

cilia

Figure 2 Examples of specialised cells (a muscle cell; b sperm cell; c nerve cell; d epithelial cell).

Science skills

Table 1 Numbers of mitochondria.

Type of human cell	Average number of mitochondria per cell (to the nearest 100)
liver	1900
kidney	1500
skin	200
small intestine	1600
muscle	1700

a What is the range in the number of mitochondria in human cells?

b Explain which method you would use to display this data.

Questions

1 Explain how it is possible to see the parts of human cells, such as the nucleus.

2 Explain how substances that enter and leave cells are controlled.

3 Why are mitochondria found in large numbers in muscle cells?

4 Ribosomes synthesise proteins. Explain what this means.

5 Explain why the nucleus of the cell is important.

6 Look at Figure 2. **(a)** Give two differences between a muscle cell and a ciliated epithelial cell. **(b)** Most cells don't move. How does each of the following help a sperm cell to move: **(i)** tail **(ii)** mitochondria **(iii)** its shape? **(c)** The epithelial cells shown help to sweep mucus containing dust out of the lungs. Describe how they are adapted for this job.

7 Compare and contrast a nerve cell and a sperm cell, both structurally and functionally.

8 Select three examples of specialised animal cells and explain how the special features of each one adapts the cell for its function.

A*

Plant and alga building blocks

Learning objectives

- describe the structure and function of the parts of plant cells
- explain how plant cells may be specialised to carry out a particular function
- compare and contrast animal and plant cells.

Route to A*

Leaves contain two different types of mesophyll cells that you can distinguish between. Figure 1 shows palisade mesophyll cells, which are adapted, by the presence of many chloroplasts, to trap light energy for photosynthesis.

Find out the name of the other type of mesophyll cells in a leaf and learn what their function is.

Plant cells

Like animal cells, plant cells usually have a cell membrane, nucleus, mitochondria, ribosomes and cytoplasm. Unlike animal cells, plant cells also have a **cell wall**, made of a carbohydrate called **cellulose**. The cellulose in the cell wall is in the form of tiny fibres. Together, these fibres are very strong so the cell wall supports the cell and strengthens it. Algae are also made of cells that have a cell wall made of cellulose. Examples of algae include seaweed and microscopic, single-celled algae that grow on tree trunks or in fish tanks, giving the water a green colour. Some plant cells also have organelles called **chloroplasts** in their cytoplasm. Inside the chloroplasts is a green pigment called **chlorophyll**. Chlorophyll is a chemical that plants use in **photosynthesis** to absorb the Sun's light energy. In photosynthesis, light energy is converted to chemical energy in the form of glucose, as food for the plant.

In the centre of many plant cells there is a large, permanent, liquid-filled space called a **vacuole**. The liquid in the vacuole is called **cell sap** and it contains sugars, salts and water. When it is full the vacuole supports the cell, making it firm. If the vacuole is less full, the cell is not so firm.

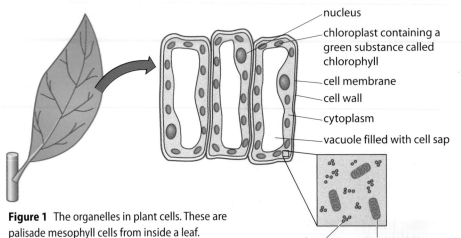

Figure 1 The organelles in plant cells. These are palisade mesophyll cells from inside a leaf.

Specialised plant cells

Many plant cells are specialised to carry out particular jobs.

- **Palisade mesophyll cells** are found in the leaf and are packed with chloroplasts. They are the main photosynthetic cells.

- **Root hair cells** have extensions into the soil to absorb water and dissolved mineral ions. These extensions are the actual root hairs. They are long and narrow so they can fit between soil particles. A thin film of water surrounds each soil particle and it contains the dissolved mineral ions.

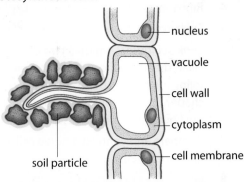

Figure 2 These root hairs are only 0.1 mm long. Each root end has thousands of these cells.

Root hairs on a germinating seed.

- **Xylem vessels** are made up of empty dead cells, arranged as long tubes of cell wall only, with no end walls between them. The cell wall has various chemicals added to it in xylem, so it does not rot away. Xylem vessels transport water from the roots, up through the stem to the leaves.

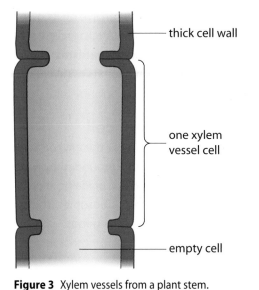

thick cell wall

one xylem vessel cell

empty cell

Figure 3 Xylem vessels from a plant stem.

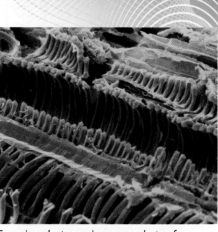

Scanning electron microscope photo of a section through a rhubarb stem, showing a xylem vessel cut open. Magnification ×290.

Science skills

Two students were measuring cells with a microscope. They used a clear plastic ruler, calibrated in mm, clipped to the microscope. The magnification of the microscope was ×10.

a Suggest how many cells they should measure to give a reliable result. Explain your answer.

b The students reported that one type of cell was 0.3 mm in diameter. Comment on the accuracy of the result bearing in mind the resolution of the ruler.

Practical

Figure 4 You can make slides of plant cells and look at them through a light microscope.

Questions

1. Why is the cell wall important in an algal cell and what is the cell wall made from?
2. Explain why plants need: **(a)** a nucleus **(b)** mitochondria **(c)** a cell membrane **(d)** ribosomes.
3. List three facts about chloroplasts.
4. What fills the large, permanent space in a plant cell? What chemicals does it contain and what is its function in the cell?
5. Explain fully the advantage of palisade cells having lots of chloroplasts.
6. Suggest which parts of a plant might not contain chloroplasts. Explain your answer.
7. Compare and contrast the structures of animal and plant cells.
8. Look at Figures 2 and 3. Explain how root hair cells and xylem vessels are adapted for their function in the plant.

A*

Examiner feedback

Distinguish carefully between the cell wall and cell membrane. A cellulose cell wall is present in plant, but not animal, cells and it allows all substances in solution to pass through it. A cell membrane is present in every cell and allows some substances to pass through, but not others. Cellulose is only found in plant and algal cells.

Bacteria and yeast cells

The most abundant cells on Earth

Tens of billions of bacteria may be present in a handful of soil. You have more bacteria in your intestines and on your skin than cells in your body. Bacteria are **unicellular organisms**. Each single cell can live on its own and carry out all the seven characteristics of living organisms. A microscope is needed to see bacteria, so they are known as **microbes** or **microorganisms**. They are found in and on plants and animals, and worldwide in habitats as diverse as deserts, deep oceans, snow and boiling mud.

The structure of a bacterial cell

Under the electron microscope we can see the internal structure of a bacterial cell. A bacterium consists of cytoplasm surrounded by a membrane and an outer cell wall. The cell wall is semi-rigid and not made from cellulose. There is no nucleus, and no other organelles, except ribosomes. The cytoplasm contains a loop of DNA that contains most of the cell's **genes**.

Examiner feedback

The mnemonic 'Mrs Gren' will help you to remember the seven characteristics of living things. They are: movement, reproduction, sensitivity, growth, respiration, excretion and nutrition.

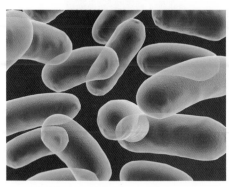

Bacteria under the electron microscope.

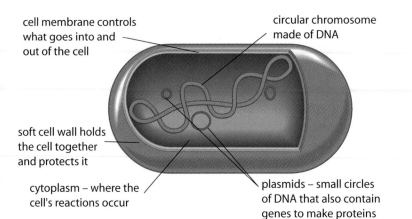

cell membrane controls what goes into and out of the cell

circular chromosome made of DNA

soft cell wall holds the cell together and protects it

cytoplasm – where the cell's reactions occur

plasmids – small circles of DNA that also contain genes to make proteins

Figure 1 Internal structure of a bacterial cell.

Taking it further

Bacterial cells have no nucleus and no other organelles surrounded by a membrane. They are known as prokaryotic cells.

Yeast cells have a membrane-bound nucleus and other organelles. They are known as eukaryotic cells. Animal and plant cells are also eukaryotic.

Science skills

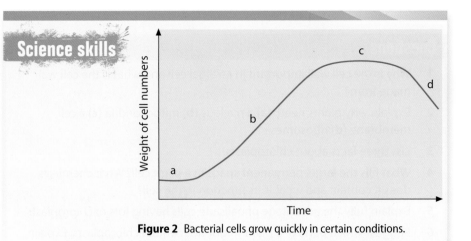

Figure 2 Bacterial cells grow quickly in certain conditions.

a Suggest which region of the graph corresponds to ideal growth conditions.

b Give a reason for your answer to part **a**.

c Suggest what is happening to the bacteria in region 'd' of the graph.

Yeast is a single-celled microscopic fungus

Some fruits – like grapes, plums and apples – often have a pale grey 'bloom' on their surface. This is partly due to naturally occurring yeast. If you have ever polished a plum or apple by rubbing it on your clothes you have removed the natural yeast. Yeast occurs on plant leaves, flowers and in the soil. Yeasts are also found in dust, water, milk and even on some of the inside surfaces of our body, such as the linings of the body cavities and various tubes.

Yeast growing on fruit skins.

Structure of a yeast cell

You can buy a block of fresh yeast for baking bread. Each square centimetre of it contains millions of individual yeast cells.

Each yeast cell is oval or spherical and has a nucleus, cytoplasm, mitochondria, a vacuole and a cell membrane surrounded by a cell wall. Yeast cells are about 10 times bigger than bacterial cells.

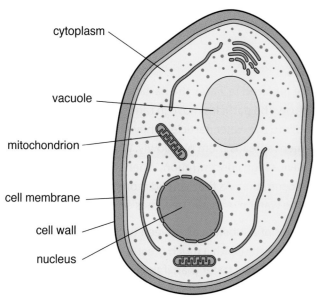

cytoplasm
vacuole
mitochondrion
cell membrane
cell wall
nucleus

Figure 3 Section through a yeast cell.

Compressed yeast. Each 1-cm³ block contains millions of yeast cells.

Questions

1 Why do we refer to bacteria as microorganisms?

2 Name and describe the only organelle found in bacterial cells.

3 Where is the genetic material found in bacterial cells and what is it made of?

4 What type of organism is a yeast cell?

5 Name and describe the organelle that contains genes in a yeast cell.

6 **(a)** Give three similarities in structure between bacterial and yeast cells.

   (b) Give two differences in structure between bacterial and yeast cells.

   (c) How much bigger are yeast cells than bacterial cells?

7 Give two ways that yeast can be used in the home and describe two forms that yeast could be in when it is bought.

8 Explain, as fully as you can, where bacteria and yeast are found naturally and why they are usually found in such large numbers.

Getting in and out of cells

Diffusion of gases

You can smell the perfume released by a flower because smelly particles spread through the air. The smell can spread many metres. The movement of the perfume particles through the particles of air is called **diffusion**.

As you get closer to the flower the smell gets stronger. This is because there are more smelly particles of gas near the flower. We say that the **concentration** of perfume particles is higher nearer the flower. The sense of smell detects chemicals in the form of a gas. Gas molecules diffuse rapidly through the air.

Science in action

In England in July 2007, smoking was banned in all enclosed public places and workplaces to prevent inhalation of smoke by non-smokers. This is called passive smoking and occurs due to diffusion of the particles in cigarette smoke through the gases in the air.

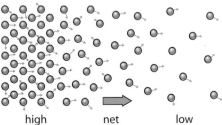

high concentration   net movement   low concentration

Figure 1 Particles move from areas of high concentration to areas of low concentration.

Route to A*

When defining diffusion, include the term 'net movement' and finish by saying that diffusion carries on until the particles in an area are equally distributed.

Diffusion of liquids

When a **soluble** substance is placed in water, the particles that make up the substance will start to diffuse. The particles move in random directions, and bump into each other and into the water particles. They start all clumped in one place, but this movement spreads them out slowly. When the particles are clumped together, they have a high concentration. When they are more spread out, they have a lower concentration. As they spread more, the concentration of the particles throughout the water eventually becomes equal.

A difference in concentrations of a substance between two areas is called a **concentration gradient**. If you start with a much greater concentration in one place than the other, diffusion will be faster than if the concentrations in the two places are nearly the same.

Practical

(a)

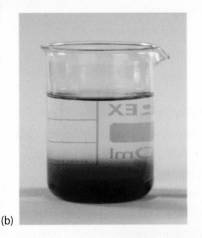
(b)

(c)

A slice of boiled beetroot is put in water at 12 noon (a). Diffusion is shown at 6 am (b) and 6 pm (c).

Diffusion through cell membranes

All plant and animal cells have a cell membrane. The cell membrane has tiny holes through which small particles can pass by diffusion. We say that cell membranes are **partially permeable membranes** because large particles cannot get through.

If the concentration of small particles on each side of a membrane is different, then more particles will diffuse through the membrane from the concentrated solution to the dilute solution. The overall or **net movement** of particles is from a higher concentration to a lower concentration. Diffusion results in an even distribution of particles on both sides of a membrane.

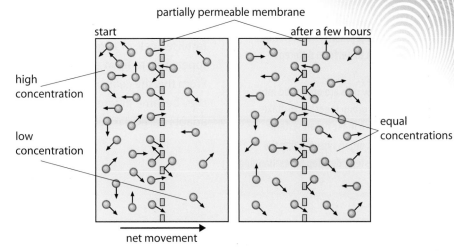

Figure 2 When the concentrations are different, the net movement of particles is from the higher concentration to the lower concentration.

All living organisms need oxygen for respiration. Oxygen molecules are small molecules that can pass through cell membranes by diffusion. There are five cell membranes for oxygen to pass through from an air sac, or **alveolus**, in the lungs, to a red blood cell that carries the oxygen round the body. The rapid rate of diffusion keeps the cells alive. It enables oxygen to get into the blood fast enough to be transported to body cells.

Questions

1 Name one sense that a butterfly uses to find a flower that is far away.

2 Explain how the perfume of the flower reaches the butterfly.

3 Where is the concentration of perfume particles highest: near the flower or far from it? Explain your answer.

4 Explain what we mean by 'partially permeable membrane'.

5 Explain how cell membranes control which particles pass through them by diffusion.

6 Explain what we mean by 'net movement'.

7 If milk is poured into a mug of hot, black coffee and not stirred, explain, using the word 'diffusion', why the coffee changes colour from black to medium brown.

8 Cells need oxygen for respiration. Explain how oxygen can get into cells.

Specialised organ systems

Differentiation from a single cell

Your multicellular body has developed from one fertilised cell, or **zygote**. This one cell divided repeatedly to form a tiny ball of identical cells. From this the cells began to **differentiate**: they became different by specialisation. This adapts particular cells for a specific function.

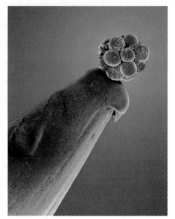

An early human embryo on a pin head, before the cells start to differentiate.

Similar cells make a tissue

Some living things are made up of only one cell. Other living things are made from millions of cells. In more complex organisms, specialised cells of the same type group together to form **tissues**. A tissue is a group of cells with similar structure and function. In animals a tissue might make a thin sheet of cells, like the epithelial cells that make up linings inside the body. Cells in other tissues group together, like muscle cells that make muscle tissue. Muscle tissue contracts to bring about movement. Cells in glandular tissue produce and release particular chemicals such as enzymes and hormones.

Plants also have tissues. Epidermal tissues cover the plant. Mesophyll tissue carries out photosynthesis in the leaf and xylem and phloem tissues transport substances around the plant.

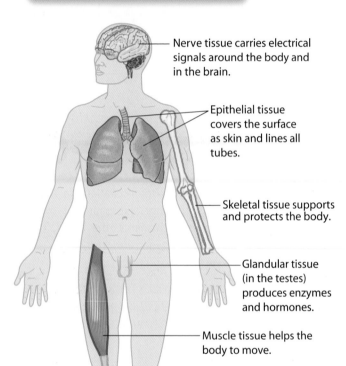

Nerve tissue carries electrical signals around the body and in the brain.

Epithelial tissue covers the surface as skin and lines all tubes.

Skeletal tissue supports and protects the body.

Glandular tissue (in the testes) produces enzymes and hormones.

Muscle tissue helps the body to move.

Figure 1 Humans have many different types of tissue.

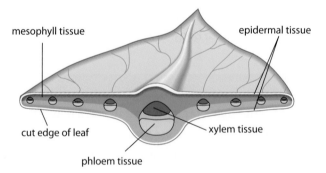

mesophyll tissue

epidermal tissue

cut edge of leaf

xylem tissue

phloem tissue

Figure 2 Section through a leaf showing some tissues.

Several different tissues make an organ

Groups of tissues join together to make more complicated structures that are called organs. For example your stomach is an organ. It has three layers of muscular tissue to churn or mix up the stomach contents as the muscles contract and relax. The inner wall contains glandular tissue that produces digestive juices. Epithelial tissue covers the outside and inside of the stomach. Other organs in your body include your heart, brain, liver and lungs. Organs have a specific function. The function of the stomach is the storage and digestion of food.

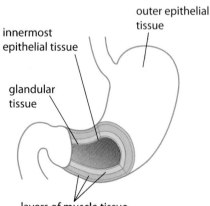

outer epithelial tissue

innermost epithelial tissue

glandular tissue

layers of muscle tissue

Figure 3 A section through the stomach's tissues.

Several different organs make a specialised organ system

Systems are groups of organs that perform a particular function. The digestive system in humans and other mammals is one example of a system where substances are exchanged with the environment. Food enters the body, is broken down and absorbed into the blood. Undigested food with some added waste chemicals is returned to the environment.

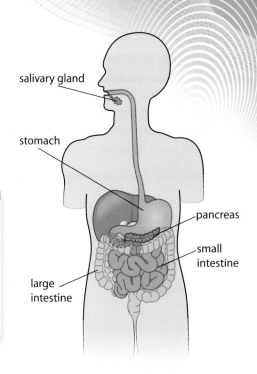

Figure 4 Some organs in the digestive system.

> **Science skills**
>
> Read the information below about the digestive system and then construct a table to show it.
>
> The digestive system includes glands such as the pancreas and salivary glands that produce digestive juices. Digestion occurs mostly in the stomach and small intestine. Absorption of soluble food occurs in the small intestine. In the large intestine, two processes occur: water is absorbed into the blood from the undigested food and faeces are produced.

Plant organs

You have just been studying the hierarchy of cell organisation in animals. You may have realised, when looking at Figure 2, that a leaf is a plant organ, as it contains several different types of tissue and has the specific function of photosynthesis in the light. Two other plant organs are roots and stems. Roots absorb water and minerals. The stem holds the leaves in a good position to catch as much light as possible. Leaves are where photosynthesis occurs in plants when they are in the light.

Questions

1 What is the function of: **(a)** epithelial tissue **(b)** glandular tissue?

2 Define an organ and write down six organs in your body.

3 Arteries are blood vessels carrying blood away from the heart. Suggest what the functions of the following tissues are in an artery: **(a)** epithelial tissue **(b)** muscular tissue.

4 Draw a table listing four plant organs and their functions. Suggest why plants need organs.

5 Outline how you could see plant cells for yourself, in a leaf you were given in the laboratory.

6 Using the stomach as an example of an organ, describe three types of tissue it contains and the function of each tissue's cells.

7 Name a system in humans where substances are exchanged with the environment. Explain how three organs are involved in the example you have given.

8 Explain the terms 'differentiate' and 'specialise', by referring to human cells and tissues.

Taking it further

Xylem, the main supporting tissue in plants, is distributed differently in stems and roots. In stems it is found as distinct oval patches arranged in a circle, while in roots it is found as a central cylinder.

Suggest what forces a plant stem and root are subject to as they grow and survive in the air and soil respectively.

How will the arrangement of xylem in the root and shoot equip them for their survival?

Route to A*

Suggest one other organ system in humans and mammals, in addition to the digestive system, where substances are exchanged with the environment.

Photosynthesis

Learning objectives

- describe photosynthesis in an equation
- describe the role of chloroplasts and chlorophyll in photosynthesis
- explain energy conversion in photosynthesis
- explain that oxygen is a by-product of photosynthesis.

A life-giving chemical reaction

The word 'synthesis' means to combine or to join together to create something new. 'Photo' means light. Photosynthesis uses light energy and two simple molecules: carbon dioxide from the air and water from the soil, to make a more complex molecule called glucose. Oxygen is also released as a **by-product**. Photosynthesis happens in a series of reactions that can be summarised by the equation:

$$\text{carbon dioxide} + \text{water} \xrightarrow{\text{light energy}} \text{glucose} + \text{oxygen}$$

Glucose is a larger molecule than carbon dioxide and water, and contains more energy in its bonds. This energy can be used for growth. Plants make glucose by photosynthesis.

Photosynthesis takes place in chloroplasts

Chlorophyll is a green pigment that is found in chloroplasts. Chloroplasts are found mainly in palisade cells in the upper layer of leaves. The chlorophyll absorbs light energy, which is used to convert carbon dioxide and water into glucose.

A variegated plant.

Some plants have green and white leaves. These leaves are called **variegated** leaves and many cultivated plants have them. Only the green parts contain chlorophyll. Photosynthesis can only take place in the green parts of the leaves.

Some algae can photosynthesise

Green algae are organisms that have some different characteristics to plants. They do, however, have chloroplasts and can photosynthesise. You may have seen green algae in ponds and streams in summer. Also, seaweeds are algae. **Plankton** contains the most numerous types of algae. They are single celled and microscopic and are found in the surface layers of lakes, rivers and oceans.

Investigating photosynthesis

One of the easiest ways to see if a plant is photosynthesising is to test it for **starch**. Any excess sugar produced during photosynthesis is stored as the insoluble product starch and this will stain blue-black with the iodine test. In the photograph of the two leaves, the brown leaf is showing the brown iodine stain because no starch is present in it. It has not photosynthesised. The other leaf is stained blue-black because it contains starch. This

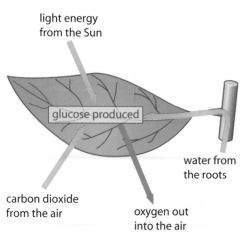

light energy from the Sun

glucose produced

carbon dioxide from the air

oxygen out into the air

water from the roots

Figure 1 Photosynthesis produces sugar.

These algae can photosynthesise.

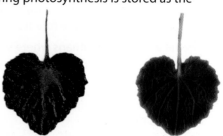
The iodine test for starch. If starch is present the leaf turns blue-black.

tells us that the leaf has photosynthesised. Before iodine is added to the leaves they are plunged into boiling water and decolourised by heating in ethanol.

You can also check that oxygen is produced in photosynthesis by collecting and testing it with a glowing splint. The easiest way to do this is by using pondweed, a plant that lives under water.

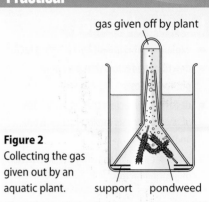

gas given off by plant

Figure 2
Collecting the gas given out by an aquatic plant.   support   pondweed

Science skills

The easiest way to measure the rate of photosynthesis is to measure the rate at which oxygen is produced. The two sets of apparatus shown in Figure 3 do this in different ways.

a   What two measurements do you need to make in order to calculate the rate at which oxygen is produced?

b   Which set of apparatus in the diagram would give the more reliable data: A or B? Explain the reason for your answer.

c   Table 1 shows results obtained from this experiment. Suggest the most suitable method of displaying the results.

Table 1   Results from experiment.

Light intensity/ arbitrary units	Rate of movement of meniscus/mm in 5 min
0	0
2	3
4	6
6	9
8	12
10	12

d   How many times should the experiment be repeated to make the results reliable?

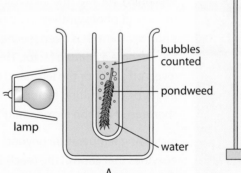

bubbles counted

pondweed

lamp

water

A

1 cm³ syringe
clamp
pondweed
dilute sodium hydrogen-carbonate solution
rubber tubing
meniscus (movement measured)

B

Figure 3   Measuring the rate of photosynthesis.

Questions

1 (a)   What are the reactants in photosynthesis?
(b) Where do they come from? **(c)** Why is light needed for photosynthesis?

2   What are the products of photosynthesis and how are they useful to the plant?

3   Explain the role of chlorophyll in photosynthesis.

4   A farmer forgets to water the crops when the weather is dry. What effect will this have on photosynthesis?

5   Describe the appearance of variegated leaves. If you had two leaves of the same type and size, that had been kept in the same conditions but one was variegated and the other was not, suggest

with reasons which would have the higher rate of photosynthesis.

6   How could you test a leaf from a plant that has been kept in a dark cupboard for 3 days to see if it had been photosynthesising? What result would you expect from the test?

7   Some tiny green algae grow on the surface of ponds. If a pond gets completely covered by these, the larger plants underneath them die. Suggest why.

8   Pondweed is a plant that lives under water. Explain how it obtains glucose.

Limiting factors

Learning objectives

- explain what is meant by a limiting factor
- describe the factors that may limit photosynthesis
- describe how factors that limit the rate of photosynthesis can interact with one another
- interpret a graph showing the interaction of limiting factors.

Crop plants photosynthesising.

What limits the rate at which crops grow?

Growing tomatoes is big business. Tomato plants produce our food through photosynthesis. Growers need to know the conditions in which photosynthesis works fastest if they are to harvest the largest possible crop.

Rate of photosynthesis

The rate of photosynthesis is the speed at which photosynthesis takes place. It is affected by the environment. The factors that affect it most are temperature, **light intensity**, availability of carbon dioxide and availability of water. The rate of photosynthesis is limited by low temperature, shortage of carbon dioxide and shortage of light.

If the level of one or more of these is low, the rate of photosynthesis will be slowed down, or limited. The factor that is reducing the rate of photosynthesis is called the **limiting factor**. A limiting factor is something that slows down or stops a reaction even when other factors are in plentiful supply.

How do variations in the amount of light, carbon dioxide and temperature affect the rate of photosynthesis?

If we are to advise growers we need to investigate how variations in these three factors affect the rate of photosynthesis. When we investigate each factor separately we find that increasing the amount of the limiting factor will increase the rate of photosynthesis, but only up to a certain value. After this rate of photosynthesis has been reached there is no further increase. Some other factor has become a limiting factor. This can be seen in Figure 1 where each graph levels off at 'X'. For the left-hand graph low temperature or low light intensity might be the limiting factor. For the right-hand graph low temperature or shortage of carbon dioxide might be the limiting factor.

Increasing temperature beyond the optimum value for a plant causes the enzymes that control the reactions of photosynthesis to break down or **denature**. This stops photosynthesis.

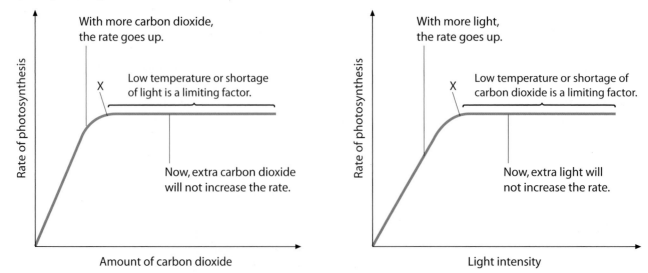

Figure 1 Carbon dioxide and light intensity affect the rate of photosynthesis.

Interaction of limiting factors

In practice, light intensity, temperature and the level of carbon dioxide interact to affect the rate of photosynthesis. Any one of them might be the limiting factor at a specific time of day. For instance at dawn and in the evening the temperature may be low, or in rainy weather thick cloud may reduce the light intensity. The Earth's atmosphere has a very low concentration of carbon dioxide of about 0.04% and carbon dioxide can be a limiting factor when plants grow densely together as crops or in tropical rainforests.

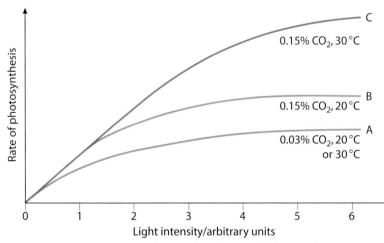

Figure 2 Interaction of limiting factors affects the rate of photosynthesis.

Questions

1   What is meant by the rate of photosynthesis?
2   List three factors that affect the rate of photosynthesis.
3   Explain what a limiting factor is.
4   **(a)** Using the three limiting factors discussed on this page, construct a table and suggest, using ticks, which of them you think may be limiting in the following locations: the Arctic, a hot desert, a tropical rainforest. **(b)** Which other substance in short supply will limit or prevent photosynthesis?
5   Explain the financial implications for a grower of ignoring the concept of limiting factors.
6   Look at the graphs in Figure 1. **(a)** Suggest why increasing the amount of carbon dioxide increases the rate of photosynthesis. **(b)** Suggest why increasing the amount of light can increase the rate of photosynthesis. **(c)** Explain why both graphs level off as the factor continues to increase. **(d)** The percentage of carbon dioxide in the air is about 0.04%. On a warm, sunny day, suggest which factor is limiting the rate of photosynthesis in the middle of a crop. Give a reason for your answer.
7   Look at Figure 2. **(a)** Which curve shows the highest rate of photosynthesis at 6 units of light intensity? **(b)** Which curve shows the lowest concentration of carbon dioxide? **(c)** Explain why curve C is much higher than curve B. **(d)** Suggest why you get curve A even if the temperature is increased from 20 °C to 30 °C.
8   You have been asked to experiment and find out the rate of photosynthesis of a specimen of pondweed provided by your teacher. Outline what you would do, giving reasons for your proposed method.

Route to A*

Curves with a similar shape to those in Figure 1 are very common in biology. You need to be able to explain why the curves go up to start with and why they then level out.

Taking it further

Plants are **autotrophic** in their nutrition – this means they feed themselves. Plants build up organic molecules from simple inorganic molecules, using light as a source of energy.

The biochemical pathway involved has two main stages: the light-dependent reaction and the light-independent reaction. Suggest which of these stages is mostly limited by temperature.

Manipulating the environment of crop plants

Propane burner for carbon dioxide enrichment, installed in a greenhouse.

Economics of enhancing photosynthesis

High energy costs for supplementary heating, lighting and carbon dioxide concentration prohibit some potential growers from using greenhouses and polytunnels. In addition there are one-off costs for building and equipment to set up this type of agriculture. However, it results in a bigger yield, and crops grown out of season usually sell for more money.

Getting carbon dioxide to greenhouse crops

Three ways of supplying carbon dioxide to greenhouse crops are described below. The information in each section is for 4 hectares of greenhouse maintaining a carbon dioxide concentration of 1300 parts per million (ppm).

Propane burners

When propane is burned, carbon dioxide is produced and heat is released. Propane is derived from fossil fuel and these fuels tend to contain sulfur as an impurity. If there is sulfur in the fuel, sulfur dioxide will also be released. About 1.4 kg of water is released for each cubic metre of propane burned. The one-off cost of installing the burners is £32 600 and the daily cost of propane is £217.

Carbon dioxide from flue gases

Natural gas is burned in a microturbine, which is used to generate electricity. The heat released during combustion is used to heat water. This can be circulated immediately throughout the greenhouse by pipes, or stored in large tanks for use at night. The carbon dioxide in the flue gas is distributed to the crops through a pipework system. The one-off cost of the equipment is approximately £118 000 and the natural gas fuel for the microturbine costs £84 per day.

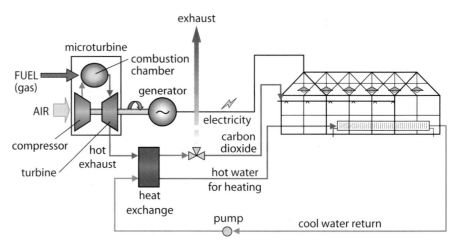

CO_2-enrichment from a combined heat and power (CHP) unit.

Liquid carbon dioxide

Liquid carbon dioxide is pure carbon dioxide. It is delivered in bulk by tankers and stored in special cylinders. The liquid carbon dioxide is vaporised then delivered to the plants by PVC tubing with a hole punched near each plant. The equipment for storing and vaporising the carbon dioxide is rented for £6900 per year and the daily cost of the carbon dioxide is £234. This method of providing carbon dioxide is used more on mainland Europe than in the UK. Once installed, equipment can be expected to last for at least 10 years.

Bulk storage of liquid carbon dioxide.

Science skills

During the day, plants both photosynthesise and respire. The relationship between gross photosynthesis, net photosynthesis and respiration is given in the equation:

gross photosynthesis = net photosynthesis + respiration

Table 1 The rates of gross photosynthesis and net photosynthesis for a cereal crop at different temperatures.

Temperature/°C	Rate of gross photosynthesis/ arbitrary units	Rate of net photosynthesis/ arbitrary units
12	12	10
19	26	24
26	40	37
34	34	27
41	26	11

a Plot a graph of the data in Table 1. Choose suitable scales for the axes. Label each of the curves.

b Describe the effect of temperature on the rate of gross photosynthesis.

c Which factor is limiting the rate of gross photosynthesis between 19 °C and 26 °C? Explain the reasons for your answer.

d The rate of gross photosynthesis is the same at 19 °C as it is at 41 °C. The cereal crop grows more slowly at 41 °C than at 19 °C. Suggest an explanation for this.

Cucumbers are now grown mainly in greenhouses.

Table 2 The yield of cucumbers grown in a well-lit greenhouse under different conditions.

Temperature/°C	Yield of cucumbers/kg per 10 plants	
	0.13% carbon dioxide	0.04% carbon dioxide
12	12	10
19	26	24
26	40	37
34	34	27
41	26	11

e In which conditions did the cucumbers give the greatest yield?

f Would the grower make most profit by using these conditions? Explain the reasons for your answer.

Questions

1 In a table, summarise the advantages and disadvantages of each of the methods of supplying carbon dioxide described above.

2 Imagine you are a grower. Which method would you use? Explain the reasons for your answer.

A*

Assess yourself questions

1 The drawing shows part of a plant as seen through an electron microscope.

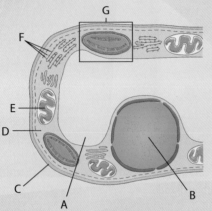

Figure 1 Part of modified plant cell.

(a) Name the structures labelled A–G. *(7 marks)*

(b) Give the function of the part labelled:

   (i) E   (ii) F   (iii) G   *(3 marks)*

(c) 1 µm is 1/1000 mm. The length of five of the structures labelled E were measured. Their lengths were as follows:

   5.1 µm   5.5 µm   5.8 µm   5.4 µm   5.7 µm

   Calculate the mean length of structure E.   *(1 mark)*

2 Figure 2 shows the structure of the type of muscle that moves our limbs.

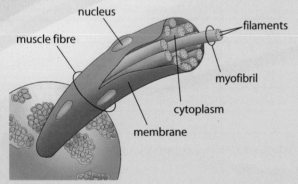

Figure 2 Muscle fibres and filaments.

(a) Give a difference between a muscle fibre and a typical animal cell related to the nucleus.   *(2 marks)*

(b) There are large numbers of mitochondria in a muscle fibre. Explain why the muscle fibre needs so many mitochondria.   *(2 marks)*

(c) (i) Suggest the function of the filaments.   *(1 mark)*

   (ii) Suggest the advantages of having many filaments in a fibre.   *(1 mark)*

3 (a) Explain what is meant by diffusion.   *(2 marks)*

(b) Figure 3 shows four ways in which molecules may move into and out of a cell. The dots show the concentration of molecules.

Which arrow, A, B, C or D, represents the movement of:

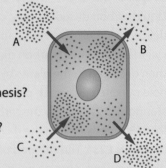

(i) carbon dioxide during photosynthesis?

(ii) carbon dioxide during respiration?   *(2 marks)*

Figure 3 Cell and molecules.

4 (a) What is meant by a limiting factor?   *(1 mark)*

(b) Figure 4 shows the effect of light intensity, carbon dioxide concentration and temperature on the rate of photosynthesis.

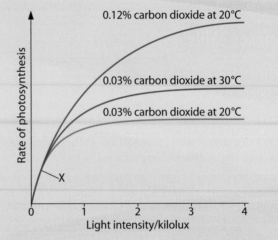

Figure 4 Light intensity versus rate of photosynthesis.

(i) Which factor is limiting the rate of photosynthesis at X?   *(1 mark)*

(ii) In a greenhouse in winter the carbon dioxide concentration is 0.03%, the temperature is 20 °C and the light intensity is 3 kilolux.

   Using the data on the graph, predict whether increasing the carbon dioxide concentration to 0.12% or the temperature to 30 °C would result in the greater increase in the rate of photosynthesis. Explain your answer as fully as you can.   *(2 marks)*

5 (a) Explain what is meant by:

   (i) digestion   (ii) differentiation

   (iii) specialisation.   *(3 marks)*

(b) Use *all* the information in Table 1 to write an educational flyer about the role of a specialised organ system like the digestive system. You must write in full sentences.

In this question you will be assessed on using good English, organising information clearly and using specialist terms where appropriate.   *(6 marks)*

Table 1 Information about the digestive system.

	Examples (random order)
Tissues	muscular, glandular, epithelial
Organs	pancreas, stomach, large intestine
Digestive functions	mix stomach contents, digestive juices, absorb water

6 A student studied ivy plants of the same species growing against a fence in her garden. She noticed that the leaves on the plants were not all the same size. She thought there might be a link between the height above the ground and the size of the leaves.

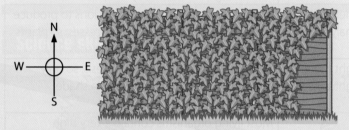

Figure 5 A fence with ivy on it.

She measured the surface area of five leaves at five different heights above the ground.

Table 2 The student's results.

Height above ground/ cm	Area/cm²					Mean surface area of leaves/cm²
	Leaf 1	Leaf 2	Leaf 3	Leaf 4	Leaf 5	
30	11	15	14	11	12	13
60	19	20	24	18	26	23
90	35	32	34	32	37	
120	44	41	40	43	40	42
150	57	43	49	52	55	51

(a) Calculate the mean surface area of the leaves collected at 90 cm above the ground. *(1 mark)*

(b) What is the range of size of leaves for 60 cm above the ground? *(1 mark)*

(c) Display the student's results as a graph. *(8 marks)*

(d) Describe how the mean surface area of the leaves is related to their height above the ground. *(2 marks)*

(e) Copy and complete the sentence by choosing the correct word from the box:

precise, reliable, valid, variable

The mean could have been improved by sampling 10 leaves instead of five. This would have made the mean more _____. *(1 mark)*

(f) The student thought that the further away from the ground the leaves were, the more light they received. How could she measure this in her garden? *(1 mark)*

(g) Suggest *two* other factors that could influence leaf size in the ivy plant. *(2 marks)*

7 A group of students was studying the distribution of daisy and dandelion plants in a field by counting along a transect.

(a) Explain what a transect is. *(1 mark)*

(b) How would sampling be done with quadrats along the transect? *(2 marks)*

(c) **Table 3** The students' results.

Distance along the transect/m	Percentage cover of plants	
	Daisy	Dandelion
5	30	10
10	20	20
15	5	5
20	5	–
25	10	5
30	20	10
35	45	15
40	40	20

Plot the results on a graph. *(6 marks)*

(d) Describe the distribution of the two plants along the transect. *(3 marks)*

(e) The students also measured soil depth at each distance along the transect and the results are shown below.

Table 4 The students' results.

Distance along the transect/m	Soil depth /cm
5	20
10	25
15	8
20	6
25	10
30	15
35	>30
40	>30

Add these data to your graph. *(4 marks)*

(f) Suggest an explanation for the distribution of dandelions along the transect. *(2 marks)*

(g) Suggest two other abiotic factors that could be measured. *(1 mark)*

Collecting ecological data

Monitoring biodiversity

The Government is planning to build a high-speed rail link between London and Birmingham. The outline route passes through countryside that has a large biodiversity. Scientists will survey the wildlife in these areas in order to report on the effects of the proposals on threatened species. Some of the techniques the scientists will use are described below.

How many woodlice live in a wood?

Imagine you want to make a count of the number of woodlice in an area of woodland. Woodlice are tiny, there are very many of them and they are not distributed evenly across their habitat. Because of this it is not easy to find out how many woodlice there are in the wood. Instead of trying to count all the woodlice, we can use **sampling**. We count the numbers in a small area and use this number to estimate the total.

Large numbers of woodlice live in damp woodland soil.

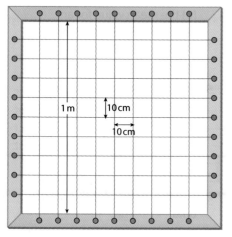

Figure 1 A 1 m quadrat divided into 10 cm squares.

Quadrats

The most common method of sampling organisms is to use a **quadrat**. This is a square frame, usually measuring 10 cm, 50 cm or 1 m along each side. Quadrats are usually subdivided into smaller squares.

Science skills

A group of four students each placed a 10 cm quadrat on the floor of a wood. Figure 2 shows their quadrats.

a Count the number of woodlice in quadrat A. Use this result to estimate the number of woodlice in 1 m² of woodland.

b Now count the total number of woodlice in quadrats A, B, C and D. Divide the total to find the mean number of woodlice in a 10 cm quadrat. Use the mean number to estimate the number of woodlice in 1 m² of woodland. What does your second estimate tell you about using quadrats?

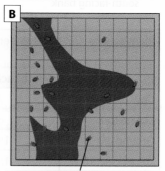

leaf litter

woodlouse

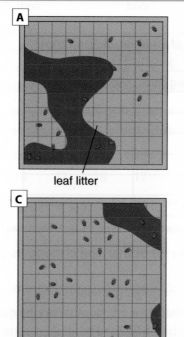

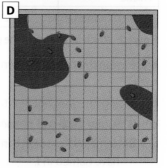

Figure 2 The students' quadrats.

Reliability and validity

The greater the number of quadrat counts that are made, the more reliable the estimate of the size of the population will be. Increasing reliability in this case also increases the validity of the population estimate.

Sampling methods

Not all parts of an area being sampled will be the same – for example, the lawn on the right may have some patches that are full of clover, and some that are almost clover-free. To get more valid results from sampling, quadrats should be placed carefully. There are two approaches to placing quadrats:

- **Random sampling**. A set of random numbers is generated by a computer. The numbers are used as coordinates on a grid as shown in Figure 3a. A similar grid is marked out on the lawn and the quadrat is used at each of the random coordinates.

- **Systematic sampling**. A grid is marked out on the lawn and the quadrat is used at each intersection, as shown in Figure 3b.

Both random and systematic sampling avoid bias in placing the quadrats.

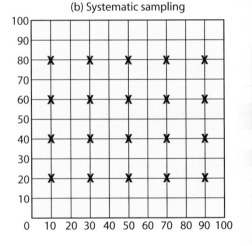

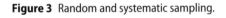

Figure 3 Random and systematic sampling.

Science skills — Quadrats can be used to estimate **ground cover** as well as for counting populations. Clover grows in clumps among grass. A student wanted to find out how much of a lawn was covered by clover. She placed a 50 cm quadrat on the lawn as shown in Figure 4.

Clover growing among grass.

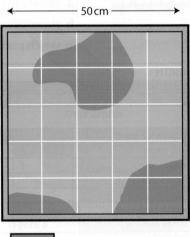

Area covered with clover

Figure 4 Using a quadrat to estimate cover.

To estimate the area of lawn covered by clover, count any square more than half-covered by clover as a 'clover square'.

c Use the number of 'clover squares' to calculate the percentage of squares covered by clover.

d How could the quadrat be modified to give a more accurate measurement of cover? Give a reason for your answer.

Questions

1. How can the reliability of results obtained using quadrats be improved?

2. How can the accuracy of results obtained by using quadrats to estimate cover be improved?

3. Explain how quadrats could be placed randomly to sample organisms in a habitat.

4. Explain how quadrats could be placed systematically to sample organisms in a habitat.

5. Look again at the quadrats labelled A, B, C and D in Figure 2. **(a)** Where are most of the woodlice found? **(b)** Suggest a hypothesis to explain this distribution. **(c)** Design an investigation to test this hypothesis.

ISA practice: earthworm distribution

A farmer has asked students to investigate how the outside air temperature each month affects the distribution of earthworms in the soil. Earthworms are important in agriculture in the recycling of nutrients in the soil. Earthworms live in burrows in soil. The depth at which they live depends mainly on physical factors in the soil.

Section 1

1 Write a hypothesis about how air temperature affects the distribution of earthworms. Use information from your knowledge of earthworm behaviour to explain why you made this hypothesis.   *(3 marks)*

2 Describe how you could carry out an investigation into this factor.

   You should include:

   - the equipment that you could use
   - how you would use the equipment
   - the measurements that you would make
   - how you would make it a fair test.

   You may include a labelled diagram to help you to explain the method.

   In this question you will be assessed on using good English, organising information clearly and using specialist terms where appropriate.   *(6 marks)*

3 Think about the possible hazards in the investigation.

   (a) Describe one hazard that you think may be present in the investigation.   *(1 mark)*

   (b) Identify the risk associated with this hazard that you have described, and say what control measures you could use to reduce the risk.   *(2 marks)*

4 Design a table that you could use to record all the data you would obtain during the planned investigation.   *(2 marks)*

   Total for Section 1: 14 marks

Section 2

Two students, Study Group 1, investigated how outside air temperature affects the distribution of earthworms in the soil. They measured the air temperature and counted the number of worms in 1 m² of soil. Their results are shown in Figure 1.

January		May		September	
3°C	22 worms	8°C	92 worms	12°C	30 worms
February		June		October	
1°C	8 worms	15°C	12 worms	9°C	52 worms
March		July		November	
2°C	12 worms	20°C	5 worms	8°C	78 worms
April		August		December	
5°C	38 worms	16°C	18 worms	6°C	52 worms

Figure 1 Study Group 1's results.

5 **(a)** Plot a graph of these results.   *(4 marks)*

   (b) What conclusion can you draw from the investigation about a link between outside air temperature and the distribution of earthworms? You should use any pattern that you can see in the results to support your conclusion.   *(3 marks)*

   (c) Look at your hypothesis, the answer to question 1. Do the results support your hypothesis? Explain your answer. You should quote some figures from the data in your explanation.   *(3 marks)*

Below are the results of three more studies.

Figure 2 shows the results from another two students, Study Group 2.

Study Group 2					
January		May		September	
3°C	22 worms	8°C	68 worms	12°C	40 worms
February		June		October	
1°C	2 worms	15°C	13 worms	9°C	48 worms
March		July		November	
2°C	6 worms	20°C	12 worms	8°C	65 worms
April		August		December	
5°C	30 worms	16°C	17 worms	6°C	59 worms

Figure 2 Study Group 2's results.

A third group of students, Study Group 3, decided that another factor might also be affecting the number of earthworms. They decided to find out the rainfall for the area. Their results are shown in Figure 3.

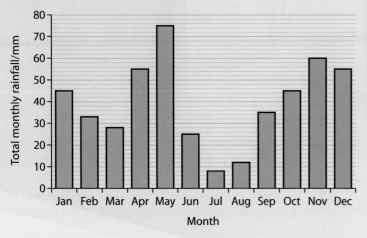

Figure 3 Monthly rainfall data from Study Group 3.

Study Group 4 was a group of scientists in India. They compared earthworm populations in two different national parks, X and Y.

- In each park they sampled eight sites by digging out a piece of soil 25 cm × 25 cm × 20 cm.
- They counted the number of earthworms in each sample.
- They measured the biomass of the worms using an electronic balance.
- They also measured the pH, nitrogen, organic matter, phosphorus, calcium, temperature and moisture content of each sample.

Table 1 shows the mean number and mean mass of earthworms from the two national parks.

Table 1 Mean number and mean mass of earthworms.

Park	Mean number of earthworms/m^3	Mean mass of earthworms / m^3/g
X	82	11.20
Y	17	2.87

Table 2 Average soil characteristics of national parks X and Y.

	Park X	Park Y
pH	6.40	6.02
Total nitrogen as a percentage	0.75	0.58
Organic matter as a percentage	5.13	4.09
Phosphorus as a percentage	0.34	0.23
Potassium as a percentage	0.90	0.63
Nitrate as a percentage	0.23	0.19
Temperature/°C	29.02	29.40
Water as a percentage	26.60	13.66

6 (a) Draw a sketch graph of the results from Study Group 2. *(3 marks)*

(b) Look at the results from Study Groups 2 and 3. Does the data support the conclusion you drew about the investigation in answer to question 5(a)? Give reasons for your answer. *(3 marks)*

(c) The data contain only a limited amount of information. What other information or data would you need in order to be more certain whether the hypothesis is correct or not? Explain the reason for your answer. *(3 marks)*

(d) Look at the results from Study Group 4. Compare the data from Study Group 1 with Study Group 4's data. Explain how far the data shown supports or does not support your answer to question 5(b). You should use examples from Study Group 4 and Study Group 1. *(3 marks)*

7 (a) Compare the results of Study Group 1 with Study Group 2. Do you think that the results for Study Group 1 are *reproducible*? Explain the reason for your answer. *(3 marks)*

(b) Explain how Study Group 1 could use results from other groups in the class to obtain a more *accurate* answer. *(3 marks)*

8 Applying the results of the investigation to a context.

Suggest how ideas from the original investigation and the other studies could be used by the farmer in encouraging growth in the numbers of earthworms in his fields. *(3 marks)*

Total for Section 2: 31 marks

Total for the ISA: 45 marks

Assess yourself questions

1 The map shows the temperatures at noon for 1 day in the UK.

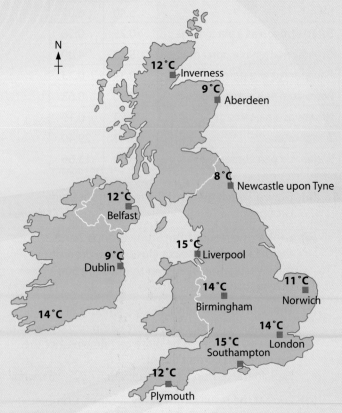

Figure 1 Noon temperatures across the UK.

Calculate:

(a) the median temperature *(2 marks)*

(b) the mode temperature. *(2 marks)*

In each case show your working.

2 The diagram shows three ways in which plant species might be distributed in a field.

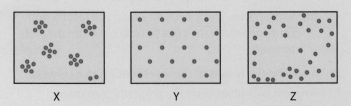

Figure 2 Three distributions of plants in a field.

(a) Describe each of the patterns X, Y and Z. *(3 marks)*

(b) At first glance a species of plant, W, appears to be distributed in a similar way to the plant species in diagram X.

   (i) Describe fully how you would investigate the distribution of species W in the field. *(3 marks)*

   (ii) Explain how your results would confirm a distribution of species W similar to that in diagram X. *(2 marks)*

3 The diagram shows some of the organisms that live in a pond.

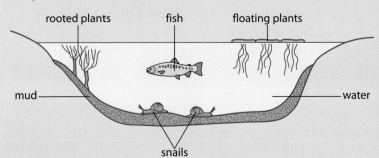

Figure 3 Organisms in a pond.

(a) Suggest *three* reasons for the different distributions of the rooted plants and the floating plants in the pond. *(3 marks)*

(b) Suggest *two* reasons for the different distributions of the fish and the snails in the pond. *(3 marks)*

4 Leafhoppers are small insects that live on leaves in trees. These insects feed on sugars, which they suck from the leaf tissues.

Students sampled the numbers of leafhoppers living on leaves in different conditions in three trees.

Table 1 The students' results.

Tree sampled	Conditions for leaves	Number of leaves examined	Number of leaves with leafhoppers on them
1	sunshine	104	13
1	shade	105	2
2	sunshine	192	44
2	shade	182	18
3	sunshine	93	19
3	shade	0	0

(a) Suggest the hypothesis the students were investigating. *(1 mark)*

(b) The students should have done further calculations on their results.

   (i) What should the students have calculated to make the results more valid? *(1 mark)*

   (ii) Copy the table then add a further column. Do further calculations and complete your table to give valid results. *(3 marks)*

(c) (i) What conclusion can be drawn from these results? *(1 mark)*

   (ii) Suggest an explanation for the distribution of the leafhoppers. *(2 marks)*

(d) What could the students have done to obtain more reliable results? *(1 mark)*

5 Students investigated the distribution of weeds in two different lawns.

Table 2 The students' results.

Weed species	Mean population density/ number of plants per m²	
	Regularly mown lawn	Occasionally mown lawn
Daisy	36.0	18.6
Dandelion	10.8	3.4
Field buttercup	1.2	10.0
Ribwort plantain	4.3	2.8
Greater plantain	0.9	1.5

(a) Describe how the data in the table could be collected. *(3 marks)*

(b) A gardening magazine recommends mowing lawns regularly to keep down weeds. Do the data in the table support this recommendation? Explain the reasons for your answer. *(4 marks)*

6 Mayflies are insects whose nymphs (immature stages) live mainly under stones in streams.

Mayfly nymphs can be sampled by disturbing the stones and collecting the nymphs in a net held downstream of the disturbed area.

Students investigated the distribution of two species of mayfly nymphs, X and Y, in different regions of a stream. They sampled 10 sites in each region.

Table 3 The students' results.

	Species X		Species Y	
	Shallow, fast-running water	Deep, slow-running water	Shallow, fast-running water	Deep, slow-running water
Mean number of nymphs per m²	2.38	12.88	24.50	6.00
Range	1–3	5–20	18–30	4–8

(a) Suggest how the students used the method described above to collect the data shown in the table. *(3 marks)*

(b) Suggest *two* reasons why the results obtained might not be accurate. *(2 marks)*

(c) Suggest *two* reasons for the different distributions of species X and Y. *(2 marks)*

7 Figure 4 is a kite diagram showing how abundant each organism is at each point on the shore – the broader the 'kite' at that point, the more abundant the organism. High tides reach as far as the shingle. All the organisms are seaweeds except *Arenicola* and *Littorina*.

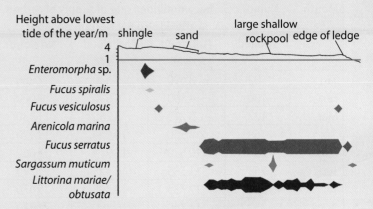

Figure 4 Distribution of some of the organisms living on a seashore.

(a) Suggest one explanation for the difference in the distribution of *Enteromorpha* and *Fucus serratus*. *(2 marks)*

(b) *Arenicola* is a marine worm. Suggest one explanation for the pattern of its distribution. *(2 marks)*

(c) *Littorina* is a snail-like organism. Suggest one explanation for the pattern of its distribution. *(2 marks)*

8 Limpets are snail-like animals that live on rocky shores.

A group of students measured the height and width of ten limpets each from a shore exposed to strong waves and a sheltered shore.

Table 4 Limpets from shore exposed to strong waves.

	1	2	3	4	5	6	7	8	9	10	Mean
Height in cm	1.7	0.9	1.4	1.7	1.2	1.9	1.7	1.2	1.9	0.9	
Width in cm	2.4	1.9	2.8	3.1	2.1	2.2	3.1	2.4	3.1	1.9	

Table 5 Limpets from sheltered shore.

	1	2	3	4	5	6	7	8	9	10	Mean
Height in cm	1.7	1.4	1.4	2.0	2.3	1.9	2.0	3.2	2.7	3.0	
Width in cm	2.8	2.1	2.8	2.3	2.6	1.9	2.5	2.6	2.5	2.9	

(a) Calculate the mean height and width of each group of limpets. *(4 marks)*

(b) Suggest an explanation for any difference in the means. *(2 marks)*

Here are three students' answers to the following question:

The table shows the recommended supply of carbon dioxide to greenhouse crops at different times of the year and in different conditions.

Month	Number of hours applied	Recommended rate of supply of carbon dioxide/kg per hectare per hour		
		Full cloud cover	Part cloud cover	No cloud
Jan	82	3 690	5 330	7 380
Feb	100	4 500	6 500	9 000
Mar	127	5 715	8 255	11 430
Apr	168	7 560	10 920	15 120
May	234	10 530	15 210	21 060
Jun	253	11 385	16 445	22 770
Jul	283	12 735	18 395	25 470
Aug	252	11 340	16 380	22 680
Sep	187	8 415	12 155	16 830
Oct	157	7 065	10 205	14 130
Nov	89	4 005	5 785	8 010
Dec	67	3 019	4 361	6 039

Suggest explanations for different recommended rates of carbon dioxide supply to greenhouse crops. *(6 marks)*

In this question you will be assessed on using good English, organising information clearly and using specialist terms where appropriate.

Read the answers together with the examiner comments. Then check what you have learnt and try putting it into practice in any further questions you answer.

Read the whole question carefully.

- Before beginning to answer a data question, jot down the main trends: in this case the seasonal trends and the sky-cover trends.
- Next, decide which concept the question is addressing: in this case limiting factors for photosynthesis. It is a good idea to introduce this concept in the first part of your answer, then to keep referring back to it as appropriate.
- If the data refer to industry, such as horticulture here, make sure you look for economic and/or environmental issues: in this case the cost of providing carbon dioxide.
- Always use the correct biological terminology.

B Grade answer

Student 1

Always refer to the rate of a process – do not simply state that a process is fast.

Light is the factor that should be referred to – it would be better to state 'there are more hours of daylight in summer'.

The recommended rate of carbon dioxide supply is low in winter and high in summer. This is because it is colder in winter than in summer so photosynthesis will not be as fast. Also the days are longer in summer. The supply is faster on days when there is no cloud. This is because there is more sun so there will be more photosynthesis.

'Sun' is far too vague. Plants receive both heat and light from the Sun – it is important to use 'light' and 'heat' rather than 'Sun'.

Examiner comment

The candidate has referred to three patterns – the change in temperature during the year, the change in day length during the year and the change in overhead conditions. These were weakly linked to photosynthesis. There is no reference to the number of hours of carbon dioxide supply. There is no reference to limiting factors or to the economics of supplying carbon dioxide.

Grade answer

Student 2

The candidate has not distinguished between light intensity and duration of light.

The candidate has correctly referred to maximum rate, but has not referred to limiting factors.

The number of hours that carbon dioxide is supplied varies with the season. This number is low in winter, rises during spring to a peak in summer, and then falls in winter. This is because day length and air temperature vary seasonally with the same pattern. Increases in light and temperature both increase the rate of photosynthesis, so the supply of carbon dioxide is increased for the maximum rate of photosynthesis. The supply of carbon dioxide is also increased on sunny days to maximise the rate of photosynthesis. But giving too much carbon dioxide would be wasteful.

Rather than referring to 'sunny' days the candidate should have referred to increased light intensity.

'Wasteful' is ambiguous – it is not entirely clear what the candidate means.

Examiner comment

A good account using correct biological terminology, such as 'rate' and 'light intensity'.

Although the candidate has referred to the maximum rate of photosynthesis this has not been linked to carbon dioxide as a limiting factor. There is an attempt at a reference to economics, but merely stating 'wasteful' is insufficient.

A* **Grade answer**

Student 3

A good opening sentence that introduces the factors that affect the rate of photosynthesis.

The candidate introduces the idea of limiting factors.

The rate of photosynthesis is affected by temperature, carbon dioxide concentration and light intensity. Any of these factors may limit the rate of photosynthesis. If one factor is limiting, then increasing other factors will have no effect on the rate. It is expensive to provide carbon dioxide to glasshouses, so the amount supplied is linked to the other limiting factors so that carbon dioxide is never the limiting factor.

The candidate links economics to limiting factors.

Crop production is also affected by day length, if no artificial light is available, since the longer the day, the longer the plants can photosynthesise.

Crop plants in greenhouses receive light and heat from sunlight. The hours of sunlight rise during the spring, are high in summer and become low again in winter. The rate of carbon dioxide supply is adjusted to correspond to the amount of heat and light being received by the plants.

Light intensity is higher when there is no cloud, so the rate of carbon dioxide supply is increased so that carbon dioxide is not a limiting factor under these conditions.

Examiner comment

An excellent answer that refers to each of the patterns in the data. The candidate has used the correct biological terminology throughout. There is a good account of limiting factors and the candidate has linked this to the cost of providing carbon dioxide.

Understanding how organisms function

This section starts by introducing proteins and their functions, both inside and outside the cells of living organisms. It then focuses on enzymes, their role as biological catalysts and their mechanism of action. Examples of enzymes, both within the body and as used in the home and in industry, are explored.

Respiration is fundamental to life, and comparison of aerobic and anaerobic respiration is covered in terms of the chemicals used and produced by these processes, and their impact on the human body during exercise.

The two types of cell division, mitosis and meiosis, are then described in relation to the types of cells produced. This leads to a discussion of stem cells, what they are, how they are produced, and the social and ethical issues raised by techniques used to produce them. Genes and alleles are introduced, leading to opportunities to use and interpret genetic diagrams to explain inheritance. Further ethical issues are explored in relation to the inheritance and treatment of genetic disorders.

The final part of this section looks at the use of fossils as evidence for the theory of evolution, exploring why some species become extinct and how new species may form.

Test yourself

1 Explain the importance of proteins in a balanced diet.

2 Explain why body temperature is controlled in humans.

3 Explain the importance of exercise for keeping healthy.

4 State the purpose of respiration in cells.

5 Describe sexual reproduction in terms of gametes and genetic variation.

6 Describe the theory of evolution via natural selection.

Objectives

By the end of this unit you should be able to:

- describe the role of enzymes as biological catalysts and explain how they work
- give examples of enzymes used in the home and industry
- compare the results of aerobic and anaerobic respiration and their roles in the body during exercise
- describe changes that happen to the body during exercise and explain their importance
- compare the outcome of sexual and asexual reproduction
- explain how stem cells are being developed to treat some conditions
- make informed judgements about the economic, social and ethical issues concerning the use of stem cells and embryo screening
- construct genetic diagrams for simple genetic crosses, including the inheritance of genetic disorders, and explain what they show
- explain how fossils can be used as evidence for evolutionary theory
- suggest ways in which species may become extinct and how new species may form.

Protein structure, shapes and functions

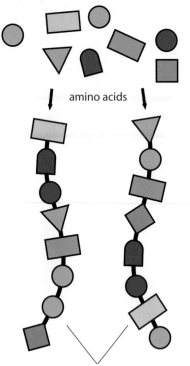

amino acids

two different polypeptides made from the same amino acids

Figure 1 Two different polypeptide chains.

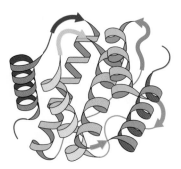

Figure 2 A folded protein contains helical (spiral) sections, and sections where the chain forms a flat zig zag (shown by arrows).

The most diverse biological molecules

Nearly 20% of the fat-free mass of your body is made up of protein – water makes up about 72%. Your hair, skin and nails are made of protein. Protein molecules, in the form of haemoglobin, carry oxygen in your blood and help if you are injured by clotting blood. Throughout the living world proteins are extremely important molecules. They take part in most chemical reactions within cells and are part of the structure of most organelles.

Proteins are structural components of muscle, hair and fingernails.

Structure of protein molecules

All proteins contain the same four elements: carbon, hydrogen, oxygen and nitrogen. Proteins are large molecules, known as macromolecules. They are made of a long chain of smaller, soluble molecules called **amino acids**. Twenty different amino acids are found in living organisms. These amino acids are joined together in a long chain, known as a **polypeptide**. Any number of amino acids can be joined together and any of the 20 types of amino acid can be used.

The sequence of the amino acids in the polypeptide chain is specific to every protein. These long chains of amino acids usually bend and fold extensively, forming a precise and specific three-dimensional shape. The shape is held together by chemical bonds. On the surface of the molecule there is often a depression or 'pocket' and this is known as the **binding site**. Other molecules can fit into the protein at the binding site.

Variety of protein functions

Some proteins form minute fibres. These have very long chains of amino acids and a simple specific shape. These give a framework or structure to some tissues such as muscles. Muscle contraction relies on proteins acting as structural components of muscle tissue.

Hormones are proteins. Insulin is a hormone that your pancreas produces. It controls your blood sugar level within narrow limits no matter what you eat. Antibodies are protein molecules with a precise 3D shape. They are produced by white blood cells to fight off invading pathogens, such as bacteria. This immune reaction helps us to survive attacks from microorganisms.

Antibodies fit and lock on to antigens that bacteria carry.

Bacteria with antigens on their surface.

White blood cells with antibodies.

Figure 3 Proteins act as antibodies.

Biological catalysts are made of protein

Thousands of chemical reactions are taking place in the cells of animals and plants all the time. The rate of these chemical reactions is increased by the action of protein enzymes, which are **catalysts**. Catalysts are chemicals that speed up the rate of reactions, but are neither reactants nor products of the reaction. As enzymes speed up reactions in living organisms they are called biological catalysts. They catalyse processes such as respiration, growth, photosynthesis and protein synthesis.

Questions

1   Why are proteins such important molecules throughout the living world?

2   How are amino acids initially arranged in protein molecules?

3   Give three reasons why the amino acid composition of proteins is so varied.

4   Describe how each protein acquires its specific three-dimensional shape.

5   Give four functions of proteins.

6   Look at Figure 1. Using the same colours and shapes to represent amino acids, draw five different polypeptide chains each with eight amino acids.

7   A man cut his chin while shaving. List at least three ways in which proteins in his body are involved in this action and the consequences of it.

8   Explain what is meant by a catalyst. Explain where biological catalysts, or enzymes, act in living organisms, giving examples of two processes in which they are involved.

Ⓐ*

Characteristics of enzymes

Learning objectives

- explain how enzyme shape is related to function
- describe the effect of temperature on enzyme action
- identify how pH affects enzyme action
- explain how the loss of 3D shape results in denatured enzymes
- interpret graphs to show the optimum conditions for the action of certain enzymes.

Molecules with a vital, special shape

During a chemical reaction the substances that are reacting are chemically rearranged as chemical bonds are broken or formed, to make new substances. We can speed up the rate of some reactions by using an enzyme. The enzyme makes it easier for the reacting substances to come together and be rearranged, so the reaction happens faster.

The starting substance of a reaction is called the **substrate**, and the substance it is converted to is called the **product**. Enzymes work by locking onto substrates. Figure 1 shows how this happens. Because of its precise shape each enzyme will only act on one type of substrate, just like a key that fits into a specific lock. When the substrate has reacted, it no longer fits the space on the enzyme, and so the products leave. This leaves the space free for more reacting substances to fit into the enzyme. The enzyme is not changed by the reaction.

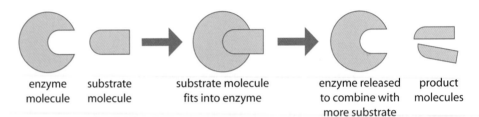

enzyme molecule   substrate molecule   substrate molecule fits into enzyme   enzyme released to combine with more substrate   product molecules

Figure 1 Some enzymes catalyse the breakdown of products.

Effect of temperature

Most chemical reactions are speeded up by an increase in temperature. Molecules move around more rapidly as the temperature rises. This causes more collisions to occur between enzymes and substrate molecules, and so increases the rate of reaction.

As the temperature continues to rise above a certain level, 37 °C in humans, the rate of enzyme-controlled reactions falls rapidly. High temperatures change the shape of enzymes: they denature them. A denatured enzyme has a different shape and so cannot lock onto the shape of the substrate. Therefore they no longer speed up the reaction.

enzyme molecule   substrate molecule

Figure 2 High temperatures change the shape of the enzyme.

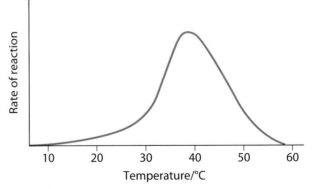

Figure 3 The effect of temperature on an enzyme-controlled reaction.

Effect of pH

Different enzymes work best at different pH values. The pH at which an enzyme works best is called the optimum pH. The optimum pH of an enzyme depends on the pH conditions where the enzyme works. For example, intestinal enzymes work best in alkaline conditions and have an optimum pH of 8. Stomach enzymes have an optimum pH of 2 because the stomach is acidic. Intestinal enzymes will not work at all in very acidic conditions.

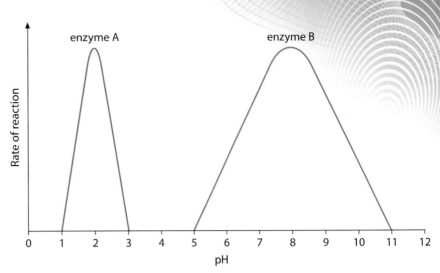

Figure 4 Different enzymes have different optimum pH values.

Potato cells contain an enzyme called catalase. This enzyme speeds up the breakdown of hydrogen peroxide. As hydrogen peroxide breaks down, bubbles of oxygen are released forming a froth. A group of students investigated how changing temperature affects the action of catalase. They were provided with potato tissue and hydrogen peroxide and were told that catalase leaves the potato across its cut surfaces. The students decided to measure the rate of reaction by recording the height of froth formed in each test tube.

Table 1 The results of the investigation.

Temperature/°C	Height of froth/cm
15	2.5
25	4.2
35	4.5
45	4.1
55	3.5

a Suggest a suitable control for this investigation.

b What is the independent variable in this investigation?

c One group of students carefully cut the potato into small discs. They used a ruler to make sure the discs were all cut to the same size. Another group of students added the same mass of potato to each test tube.

  i Why is it necessary to add the same amount of potato to each tube?

  ii Which method is the more accurate – measuring the size or the mass? Give reasons for your answer.

Questions

1 Describe what would happen if we did not have enzymes in our bodies.

2 Explain why the shape of an enzyme affects the way it works in a reaction.

3 High temperatures destroy the shape an amino acid chain makes. **(a)** What effect would high temperatures have on an enzyme-controlled reaction? **(b)** Explain your answer.

4 How does pH affect the rate of enzyme action?

5 **(a)** Make a list of four reactions that you know are catalysed by enzymes. **(b)** Describe two processes in a plant that would be affected if it contained no enzymes.

6 Look at the results in Table 1. Explain the rate of reaction when the temperature increases from: **(a)** 15 to 35 °C **(b)** 35 to 55 °C.

7 Explain why a graph of temperature versus rate of reaction for two human digestive enzymes usually shows a single curve but a graph of pH versus reaction rate of two different digestive enzymes may have two separate curves.

8 Explain as fully as you can why enzymes are sometimes described as a lock that substrates, acting as a key, fit into.

Digestive enzymes

Using the food you eat

Your food contains proteins, starches and sugars, and fats and oils. The molecules of proteins, starch and fats are huge – much too large for you to absorb into your body. This means they have to be **digested**, broken into smaller molecules, in your gut. These digestion reactions need to be quick so that you can absorb what you need from your small intestine, before the remains pass out of your body. All of these digestive reactions are catalysed by enzymes to speed them up.

Science skills

Read the information below about food transit time in the gut.

After a meal the time taken for food to travel through your gut depends on many factors. Roughly, it takes 2.5–3 hours for 50% of stomach contents to empty into the intestines. Total emptying of the stomach takes 5–6 hours. Then 50% emptying of the small intestine takes 2.5–3 hours. Total emptying of the small intestine takes 5–6 hours. Finally, transit through the large intestine takes 30–40 hours.

a Tabulate the data in the paragraph above and include total transit time for each region of the gut. Then graphically display the data for total transit time.

Soon to be catalysed by enzymes.

Enzymes work outside cells that produce them

Your gut is simply a hollow tube of different diameters and wall types. Digestive enzymes are produced by specialised cells in glands and tissues lining the gut. They are made inside cells, but they move out of the cells into the gut where they work as they come in contact with food molecules. Some mix with the food in the gut, others remain attached to the outside of cells in the gut wall. They are **extracellular** enzymes.

cells that produce digestive enzymes

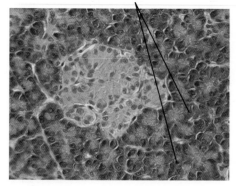
Pancreatic cells produce digestive enzymes.

The right tools for the job

Each type of food needs a particular enzyme to break it down into products that are useful to the body. Different enzymes work on different substances.

- **Amylase** enzymes catalyse the breakdown of starch to sugars.
- **Protease** enzymes catalyse the breakdown of protein into amino acids.
- **Lipase** enzymes catalyse the breakdown of lipids (fats and oils) into fatty acids and glycerol.

Type of food	Enzyme	Products

A protein molecule is made up of many different amino acids.

protease → amino acids

A starch molecule is made up of many glucose molecules.

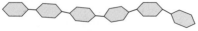

amylase → glucose

A fat molecule is made up of fatty acid and glycerol molecules.

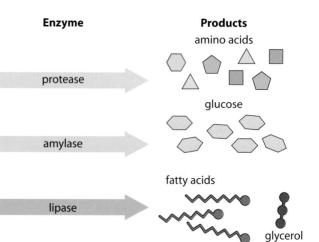

lipase → fatty acids, glycerol

Figure 1 Specific enzymes break food down into products useful for your body.

Where are these digestive enzymes produced and where do they work?

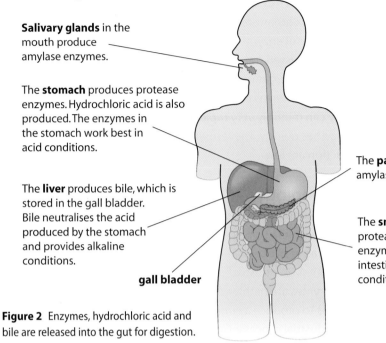

Salivary glands in the mouth produce amylase enzymes.

The **stomach** produces protease enzymes. Hydrochloric acid is also produced. The enzymes in the stomach work best in acid conditions.

The **liver** produces bile, which is stored in the gall bladder. Bile neutralises the acid produced by the stomach and provides alkaline conditions.

gall bladder

The **pancreas** produces protease, amylase and lipase enzymes.

The **small intestine** produces protease, amylase and lipase enzymes. Enzymes in the small intestine work best in alkaline conditions.

Figure 2 Enzymes, hydrochloric acid and bile are released into the gut for digestion.

Controlling gut pH

In each region of the gut other substances are released to control the pH. For example, the stomach produces hydrochloric acid because the protease enzymes there work best in an acid solution. The liver produces **bile**, which is stored in the gall bladder before being released into the small intestine. Bile neutralises the acid that was added to food in the stomach and provides alkaline conditions for enzymes in the small intestine. These are other protease, lipase and amylase enzymes.

Table 1 Each part of the digestive system releases different enzymes to digest food.

Type of enzyme	Where is it produced?	pH conditions for enzymes	Where does it work?
protease enzymes	stomach	acid	stomach
	pancreas small intestine	alkaline	small intestine
amylase enzymes	salivary glands	slightly alkaline	mouth
	pancreas small intestine	alkaline	small intestine
lipase enzymes	pancreas small intestine	alkaline	small intestine

Questions

1 There are many kinds of enzyme in your gut. Explain why.

2 What would happen if you didn't have enzymes in your gut?

3 Look at Figure 2. **(a)** In which two parts of the gut is amylase made? **(b)** Protease enzymes are made in the small intestine wall. Where else are they made? **(c)** In which part of the gut does lipid digestion take place?

4 Look at Figure 2. The stomach makes a chemical that is not an enzyme. **(a)** What is the chemical? **(b)** Suggest the conditions that enzymes in the stomach work best in.

5 You eat a cheese sandwich. Write down the stages of digestion of the starch and fat as they pass through your gut.

6 **(a)** What kind of tissue produces enzymes in general and what organs in the gut produce enzymes? **(b)** Why are digestive enzymes described as extracellular?

7 **(a)** Describe how bile reaches the small intestine. **(b)** What is one role of bile in digestion?

8 What is meant by digestion and why is it necessary in humans?

Enzymes used in industry

Learning objectives

- explain why microorganisms are useful in enzyme technology
- describe how three types of enzyme are used in the food and drinks industry
- describe some industrial uses of enzymes
- evaluate the uses of enzymes in industry.

Enzymes from microorganisms

Vast amounts of microorganisms, such as bacteria, are grown in industry to supply us with useful enzymes. Many of the enzymes produced by microorganisms are passed out of the cell, which enables scientists to use the enzyme from the microorganism. Microorganisms can be grown relatively cheaply inside vats, known as fermenters, and no expensive equipment is needed. As they can multiply very rapidly, microorganisms produce large amounts of enzymes quickly. The fermenter is kept at 28 °C so energy costs for enzyme production are low.

Enzyme technology

For many years, inorganic catalysts (those not containing carbon) have been used in industrial reactions: for example, iron is used in the production of ammonia in the **Haber process**. Recently there have been many developments in the use of enzymes as industrial catalysts, a process known as **enzyme technology**. Enzymes are highly efficient catalysts: only a small amount of enzyme is needed to produce a large quantity of product. This is why they are more useful in industrial processes than inorganic catalysts.

Enzymes in the food and drinks industry

Enzymes are now used a lot in the catering industry. In some baby foods, proteases are used to help 'pre-digest' the proteins. This makes the food softer, or less fibrous, for babies to eat and easier for babies to digest.

Millions of bacteria can be grown quickly in these fermenters. Enzymes extracted from the liquid are used for many purposes.

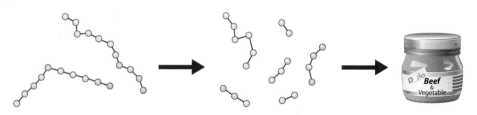

Figure 1 Protease enzymes are used to break down long-chain protein molecules into short chains that are easier for babies to digest.

Enzymes are also used in the production of sugar syrups used as sweeteners in the food and drinks industry. It is cheaper to get starch than sugars from plants to use in our food. Starch can be obtained from potatoes and cereals that are cheaper than sugar sources like sugarcane and sugar beet. Starch can be converted to sugar syrup using **carbohydrases** to catalyse the reaction. This means we can make products like sugary drinks, cakes and sweets more cheaply.

Fructose and glucose are both sugars with the same energy value. Isomerase is an enzyme used to convert glucose syrup into fructose syrup. Fructose is a much sweeter sugar than glucose, so less needs to be added to foods and drinks to make them taste sweeter. This is very useful in the production of slimming foods and low-calorie drinks.

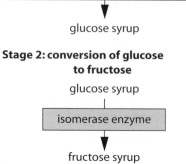

Stage 1: production of glucose syrup

starch from maize grains

Carbohydrase enzymes are added to the starch to digest it into glucose.

glucose syrup

Stage 2: conversion of glucose to fructose

glucose syrup

isomerase enzyme

fructose syrup

Figure 2 Carbohydrase and isomerase enzymes are used in the production of sweeteners.

The amount of fructose produced is affected by how fast the glucose syrup flows through the reactor containing immobilised enzymes. Figure 3 shows the result of increasing the rate of flow of glucose syrup into the reactor.

a What rate of flow should scientists use in the reactor? Explain your answer.

b How much more fructose is produced when the rate of flow is increased from 3 to 4 dm^3/min?

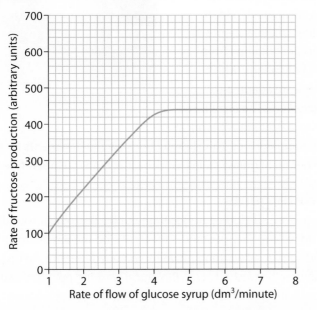

Figure 3 Rate of flow of glucose syrup versus rate of fructose production.

Questions

1 Which organisms produce the enzymes obtained in industry?

2 Where are these organisms grown and how are the enzymes obtained?

3 Give two advantages to industry of using enzymes.

4 What do the following enzymes digest: **(a)** protease **(b)** carbohydrase **(c)** lipase?

5 **(a)** Give three examples of the use of enzymes in the food industry.
   (b) For each example, name the type of food the enzyme works on and the reason for using an enzyme in the food production.

6 Evaluate the use of enzymes in the food industry.

7 Explain fully how protease enzymes in the food industry could be used over and over again to break down large protein molecules into smaller ones.

8 Evaluate the use of immobilised enzymes in the food industry.

A*

Science in action

When a reaction is complete, the enzyme and product are mixed up with each other. It is very expensive for an industry to keep producing enzymes and to keep separating products from enzymes. To avoid this expense scientists have found ways of fixing enzymes to the surface of small beads. This is called immobilising the enzyme. The diagram shows how isomerase enzymes fixed in this way can be used over and over again. It also means that there are no enzyme molecules mixed with the product, as they are all trapped in the beads.

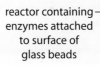

Figure 4 Immobilised enzymes can be used over and over again.

Taking it further

Isomers are molecules that contain the same types and numbers of atoms, but arranged in different ways.

Home use of enzymes

Biological detergents contain enzymes.

Science in action

Recently, manufacturers and retailers have tried to show they care for the environment by recommending that many garments are washed at 30 °C. However, Korean scientists have shown that washing at 30 °C is much worse at removing dust mites and pollen than washing at 60 °C. Allergy sufferers could have a problem with low-temperature washes.

Examiner feedback

An 'evaluate' question requires you to give some advantages and disadvantages and then to make a conclusion. Your conclusion should say whether or not you think enzymes (in this case) are useful, linked to a reason for your judgement.

Enzymes from bacteria

Bacteria help to clean our laundry. This is true if you use biological detergents containing enzymes. The first biological detergents that were produced would only work in warm water. However, proteases and lipases have now been produced that work at much higher temperatures. Most of the enzymes in washing powders are obtained from bacteria living in hot springs, which means the bacteria are adapted to live in water above 45 °C. The enzymes obtained from these bacteria will work at moderately high temperatures. This is useful because the detergents in washing powders, which get rid of greasy stains, work best at higher temperatures.

The enzymes produced by bacteria living in hot springs will work at high temperatures.

Biological detergents

The dirt that we get on our clothes comes from our bodies, our surroundings and from the food we eat. The substances that make up the dirt are mostly proteins, fats and sugars. All washing powders contain detergents to dissolve stains so that they can be washed away. Biological washing powders also contain protease and lipase enzymes. Protease enzymes catalyse the breakdown of proteins present in stains such as blood, grass and egg. Lipase enzymes catalyse the breakdown of lipids in stains such as fat, oil and grease. The protein and fat molecules in stains are broken down into smaller, soluble molecules that dissolve easily in water and can be washed away. In the home, biological detergents are more effective at low temperatures, such as 30 °C, than other types of detergents. For a wash at 60 °C or 90 °C biological detergents are not recommended because most enzymes are denatured at high temperatures and stop working. Some people still use non-biological detergents because they get an **allergic reaction**, such as a skin rash, to the enzymes.

Evaluating enzyme use in home and industry

It helps you to decide if using enzymes is useful or not by looking at their advantages and disadvantages.

Table 1 Advantages and disadvantages of enzymes.

Advantages	Disadvantages
Enzymes bring about reactions at normal temperatures and pressures, saving on energy.	Most enzymes are denatured at high temperatures.
Enzymes save on expensive equipment.	Many enzymes are expensive to produce.
Biological detergents are more effective at low temperatures than other types of detergent.	Allergy sufferers may be allergic to the enzymes and/or require a higher temperature wash to destroy allergens in bedding and clothes.

A group of students carried out an investigation to find the conditions in which biological washing powders work best. The students used photographic film to demonstrate the action of the washing powders. The film contains black grains stuck on by a layer of gelatin. Gelatin is a protein. When the gelatin is broken down by the enzymes in the washing powder the film becomes clear as the black grains come away.

The students prepared a 1% solution of washing powder by dissolving 1 g of powder in 100 cm^3 of water. Figure 1 shows how the students designed the investigation.

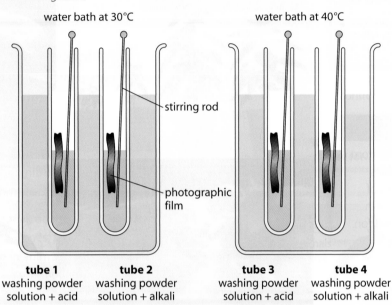

Figure 1 A biological washing powder can dissolve the protein on photographic film.

a Identify the two independent variables.

b What type of variable is the dependent variable?

c **i** Explain why a 1% solution of washing powder was used in all four test tubes.

  ii Explain why a stirring rod was used.

d Suggest what the students should do to make their results more reliable.

Table 2 The students' results.

	Tube 1	Tube 2	Tube 3	Tube 4
Temperature/°C	30	30	40	40
pH	4	8	4	8
Time taken for film to go clear/min	not digested	25	40	10

Questions

1 Which organisms produce the enzymes found in biological detergents?

2 Which enzymes digest: **(a)** protein **(b)** fat?

3 Suggest two stains that: **(a)** protease **(b)** lipase would act on.

4 Suggest: **(a)** two advantages of using enzymes to help get clothes clean **(b)** one disadvantage of using enzymes in biological detergents.

5 Explain why it is easier for stains to leave clothes when they have been digested by enzymes.

6 A student washed his clothes using biological detergent on a very hot wash at 90 °C. They still came out with stains on. Explain to him the advantages of using enzymes in detergents and why his clothes would have been washed cleaner if the water temperature had been at 40 °C.

7 Suggest two reasons why hotels do not use biological detergents to wash their cotton sheets.

8 Use Table 1 to evaluate the use of enzymes in home and industry.

Anaerobic respiration

Taking it further

You will find that advanced texts talk about lactate rather than lactic acid. This is because when lactic acid is in solution in water, it dissociates into negative lactate ions and positive hydrogen ions. However, the two terms mean the same thing in this instance.

Examiner feedback

A different form of anaerobic respiration occurs in many microorganisms, including yeast. It is commonly called fermentation and its end-product is ethanol (ethyl alcohol), not lactic acid as in animal cells. It is important to distinguish between the two processes because of their different end-products.

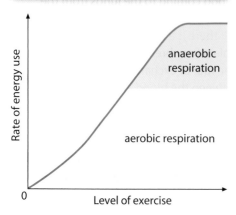

Figure 1 This graph shows the contribution of aerobic and anaerobic respiration to energy production at different levels of activity.

Running out of oxygen

If you exercise for a long time, your muscles start to **fatigue**. This means that they don't contract as strongly as they normally do, and cannot do as much work. You feel an increasing weakness and pain or cramps in the muscles.

The cause of fatigue is not well understood because there are many changes happening in muscle during activity. In prolonged activity some chemicals needed for reactions start to run out, and others that are made during activity build up.

This athlete has muscle fatigue after running a marathon.

Another source of energy

Most of the time muscles get the energy to contract from aerobic respiration. However, if you suddenly start exercising vigorously, or if you exercise vigorously for some time, your muscle cells may not be able to get enough oxygen to keep contracting hard.

Fortunately, if oxygen levels in muscle cells are low, the cells can also use **anaerobic respiration**. This process releases energy without the need for oxygen to break down glucose. Anaerobic respiration does not replace aerobic respiration. It provides muscles with extra energy beyond what they can get from aerobic respiration.

Comparing the two types of respiration

Anaerobic respiration also breaks down glucose, but it does not make the same products as aerobic respiration. The equation for anaerobic respiration is:

glucose ⟶ **lactic acid**   energy given out

Anaerobic respiration produces much less energy per glucose molecule than aerobic respiration. This is because the glucose is only partly broken down and there is still a lot of energy locked in the bonds of the lactic acid molecules. However, the breakdown of glucose to lactic acid is much faster than the breakdown of glucose to carbon dioxide and water, so anaerobic respiration can supply energy quickly.

Science skills

Studies of human athletes show that different sports depend on different combinations of aerobic and anaerobic respiration.

Table 1 Types of respiration for different activities.

Activity	Type of respiration
short-distance sprint	mostly anaerobic
middle distance, e.g. 400 m run	anaerobic and aerobic
long distance, e.g. marathon	mostly aerobic

a Explain why different kinds of athletes need to train differently to improve the efficiency of their muscle cells to manage aerobic or anaerobic respiration.

The oxygen debt

For a while after exercise, we continue to breathe deeply even though our muscles have stopped working as hard. The extra oxygen that our bodies need after exercise is called the **oxygen debt**.

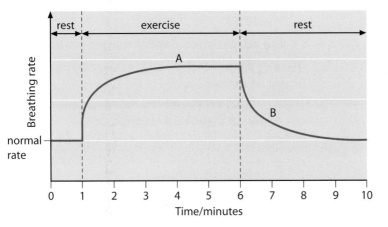

Figure 2 Breathing rate during and after exercise.

Some of the extra oxygen taken in at this time is used to return the body to its resting state. However, if anaerobic respiration has occurred, the lactic acid must also be removed from the muscle cells and recycled. It is transported in the blood to the liver where it is oxidised so that it can be used for aerobic respiration another time, when it is broken down to carbon dioxide and water.

Whales and seals keep lactic acid inside their muscles and out of their blood during diving because they dive for so long and the amount of lactic acid produced could damage other organs.

Science in action

It used to be thought that lactic acid caused muscle fatigue, but scientists now realise that it plays an essential part in keeping muscles working when they are being overstimulated during vigorous exercise. Endurance athletes are now trained to develop both their aerobic and anaerobic capacity.

Examiner feedback

Make sure you know what anaerobic respiration is, how it can contribute during exercise and why it results in an 'oxygen debt'.

Questions

1 Explain what we mean by 'muscle fatigue'.

2 State one similarity and one difference between aerobic and anaerobic respiration.

3 Explain why the oxygen concentration in muscle cells may be low during vigorous activity.

4 Give one disadvantage and one advantage of anaerobic respiration compared with aerobic respiration.

5 Look at Figure 2. Explain the breathing rate at points A and B.

6 Why is it an advantage for an endurance athlete to develop both aerobic and anaerobic capacity?

7 Explain fully why whales and seals need a special adaptation to cope with lactic acid.

8 Explain the different proportions of aerobic and anaerobic respiration shown in Table 1.

Assess yourself questions

1 Proteins are the most diverse biological molecules. Write three or four sentences about proteins. Include the following words in your sentences: polypeptide, specific, amino acids, bonds, enzymes, twenty, three-dimensional.

In this question you will be assessed on using good English, organising information clearly and using specialist terms where appropriate. (6 marks)

2 An investigation was carried out to find out the effect of bile on the action of the enzyme lipase.

Table 1 Four test tubes and their contents.

Test tube	Contents
1	milk, lipase, pH indicator, bile
2	milk, lipase, pH indicator
3	milk, boiled lipase, pH indicator, bile
4	milk, lipase, pH indicator, boiled bile

Bile is alkaline. The pH indicator is yellow when the pH is 7 or less and red when the pH is over 7.

Table 2 Time taken for the indicator to change colour.

Test tube	Time taken for the pH indicator to change colour/min
1	15
2	38
3	no change
4	15

(a) What colour was the indicator in test tube 1 at the start of the investigation? Explain your answer. (2 marks)

(b) Explain why the action of lipase caused the indicator to change colour in test tube 1. (3 marks)

(c) Explain why there was no colour change in test tube 3. (2 marks)

(d) What do the results from test tubes 1 and 2 tell you about the effect of bile on the reaction? (1 mark)

(e) One student concluded that bile contains enzymes that digest fats. Which of the results shows that this conclusion is incorrect? Explain your answer. (2 marks)

3 Scientists working for a washing powder manufacturer carried out tests on a new protease enzyme that removes protein stains, such as egg and blood. They wanted to find out if the protease was suitable for use in washing powders.

They placed equal-sized cubes of egg white into test tubes. The test tubes were placed into water baths and kept at different temperatures from 0 to 60 °C. The same volume of the enzyme was added to each tube. The scientists recorded the time taken for the egg white to be digested. Figure 1 shows their results.

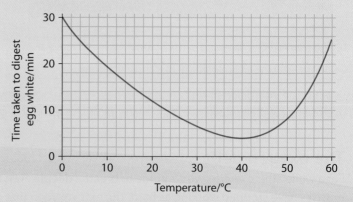

Figure 1 Results of the investigation.

(a) How long did it take to digest the egg white at 20 °C? (1 mark)

(b) Explain why the scientists used equal-sized cubes of egg white. (1 mark)

(c) How does the rate of digestion change:
  (i) between 5 and 40 °C (1 mark)
  (ii) between 40 and 60 °C? (1 mark)

(d) Is this new protease suitable for use in washing powders? Use the results of this investigation to explain your answer. (2 marks)

4 The graph below shows the action of an enzyme measured at different temperatures.

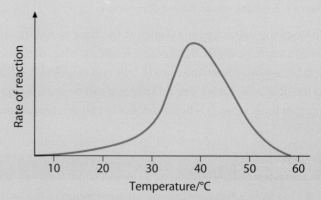

Figure 2 Enzyme action at different temperatures.

(a) Which variable is the dependent variable? (1 mark)

(b) What kind of variable is the independent variable? (1 mark)

(c) How could you show the graph was accurate? (1 mark)

(d) Explain the rate of enzyme action between 10 and 50 °C. (3 marks)

5 **Table 1** Some industrial uses of enzymes.

Use	Enzyme involved	Explanation
tenderising meat	protease	**(i)**
making sugar syrups	**(ii)**	starch is broken down into sweet sugars
(iii)	isomerase	fructose is a very sweet sugar so less of it is needed in the production of slimming foods

(a) From where are these industrial enzymes obtained?
(1 mark)

(b) What would be the consequence of the absence of enzymes from one of these chemical reactions that uses them? *(1 mark)*

(c) Complete **(i)**, **(ii)** and **(iii)** in the table above.
(3 marks)

(d) Explain what immobilised enzymes are, and how using them in reactions saves money for the food industry. *(3 marks)*

6 Respiration is a set of chemical reactions that use catalysts.

(a) Where in a cell does respiration take place?
(1 mark)

(b) Copy and complete the word equation that summarises the reactions in respiration:
glucose + _____ ⟶ _____ + _____ + energy given out *(3 marks)*

(c) To what group of chemicals do the catalysts belong and why are they important? *(2 marks)*

(d) Give *two* ways in which the energy released during respiration is used in animals. *(2 marks)*

(e) Give *two* ways in which the energy released in respiration is used in plants. *(2 marks)*

7 A student measured her heart rate immediately after sitting, walking or jogging for 2 minutes. She carried out each test three times, and rested for 2 minutes between each test. Her heart rate is given in beats per minute.

Table 2 The effect of exercise on the student's heart rate.

Exercise	Sitting			Walking			Jogging		
	1	2	3	1	2	3	1	2	3
Heart rate/ bpm	71	68	72	91	85	84	113	110	119

(a) Explain why the student took three measurements for each level of exercise. *(1 mark)*

(b) Calculate the mean for each level of exercise. *(1 mark)*

(c) Explain what the student's results show. *(1 mark)*

(d) Another student's results for the same investigation were sitting 62, walking 84, and jogging 108. Give *one* reason for the difference between the two students' results. Explain your answer. *(1 mark)*

(e) Do the second student's results support your conclusion in part **(c)**? Explain your answer. *(1 mark)*

(f) Explain fully why heart rate changes with level of exercise. *(3 marks)*

8 The graph in Figure 3 shows the different sources of energy used by muscle cells as the effort put into exercise increases.

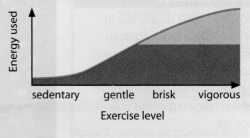

■ energy from anaerobic respiration
■ energy from aerobic respiration

Figure 3 The contribution of aerobic and anaerobic respiration in supplying energy to muscles during activity.

(a) Explain why the energy from aerobic respiration levels off as effort increases. *(1 mark)*

(b) Write a word equation for anaerobic respiration. *(2 marks)*

(c) Anaerobic respiration produces less energy per glucose molecule than aerobic respiration. Explain why. *(1 mark)*

(d) Describe what happens to the product of anaerobic respiration after exercise has ended. *(2 marks)*

(e) Explain what is meant by the term 'oxygen debt'. *(2 marks)*

(f) Explain fully why lactic acid concentration increases in muscles during vigorous activity and why it is difficult to prove that lactic acid is the cause of pain in muscles. *(4 marks)*

9 Describe fully the role of aerobic and anaerobic respiration during exercise in humans.

In this question you will be assessed on using good English, organising information clearly and using specialist terms where appropriate. *(6 marks)*

Differentiated cells

Learning objectives

- describe what cell differentiation is and when it happens in animals and plants
- explain what stem cells are and give examples of how they might be used to treat some conditions
- identify and evaluate some of the social and ethical issues surrounding the use of embryonic stem cells.

Taking it further

The process of differentiation (specialisation) involves the switching on and off of genes as the cell develops. This is why all cells have the same genes, but don't all look the same.

Science in action

Scientists are currently researching how specialised cells can be returned to the stem cell state and then directed to produce new, different specialised cells. The success of this process will change the scientific problems that need tackling to produce effective treatments for people with disorders caused by defective cells, as well as change the ethical issues raised. Society needs to be aware of any changes, and the questions that politicians and the general public will need to consider.

Differentiated cells

Your body contains many different kinds of cells, and almost all of them are **differentiated** (specialised) to do different jobs. When an animal egg cell is fertilised it starts dividing to make an **embryo**. The cells in the early stages of an animal embryo can differentiate to form almost any kind of cell. As cell division continues, the cells become increasingly specialised and the range of types of cell they can develop into decreases. By the time you are born, almost all your cells are differentiated.

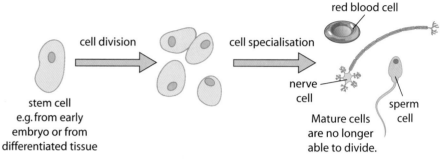

Figure 1 Differentiation of animal cells.

In plants, it is different. Many cells in a fully grown plant keep the ability to differentiate into any kind of cell. We can take cuttings from plant parts or use tissue culture to grow a whole new plant, but we cannot do this in animals.

Figure 2 In a plant cutting, cells at the cut end divide and differentiate into root cells to form a new root system.

1 Cut shoot from a plant.

2 Dip cut end into hormone rooting powder.

3 Plant in pot of soil as soon as possible after cutting.

Stem cells

Cells that can differentiate into a range of other cells are called **stem cells**. The cells in an early embryo (**embryonic stem cells**) have the ability to differentiate into any kind of body cell.

In differentiated body tissues there are a few **adult stem cells**, which divide when needed for growth or repair to damaged tissue. However, these cells can normally only differentiate into a limited range of other cells. Stem cells in **bone marrow**, for example, normally only produce different kinds of blood cell.

Using stem cells for treatment

Stem cells are already being used to treat some human disorders, for example bone marrow cells are used to treat leukaemia, a type of cancer in blood cells. New treatments are being developed for many other disorders where cells are damaged, such as to replace damaged nerve cells in people who have been paralysed after their spinal cord was broken in an accident. The advantage of using embryonic stem cells is that they are easier to work with, but a practical disadvantage is that putting cells from one body into another means lots of medication for the rest of the patient's life. Using stem cells created from a patient avoids this problem.

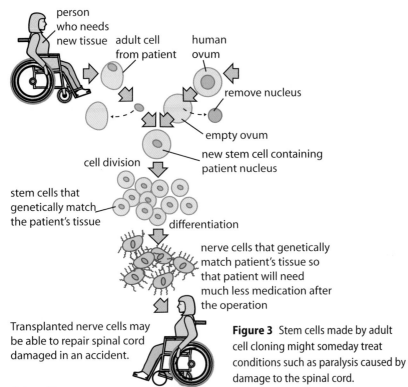

Figure 3 Stem cells made by adult cell cloning might someday treat conditions such as paralysis caused by damage to the spinal cord.

Science in action

Scientists are researching many other ways to encourage adult stem cells to differentiate into a greater range of specialised cells. For example, in March 2010 a boy was given a replacement windpipe with stem cells taken from his own bone marrow. The stem cells were treated to grow into all the kinds of cells needed in the new windpipe. Doctors will monitor the success of this new treatment.

Is it ethical?

Many countries ban the use of embryonic stem cells for treatment. This is partly because research is still trying to solve some of the practical problems they cause, such as the need for lifetime medication. However, there are also **ethical questions**. Some people say that using such embryos is destroying a new life. However, most of the embryos used are those unwanted after fertility treatment and would be destroyed anyway. Research into embryonic stem cells is allowed in many countries because they are easier to extract from tissues than adult stem cells.

Examiner feedback

You will not be expected to remember the details of Figure 3, but it will help you to understand the link between techniques for adult cell cloning and gene therapy. It also shows why people who are against the use of embryos in research and treatment don't like this technique, because an embryo is created and then destroyed.

Questions

1. Explain why we can grow a whole new plant from a leaf but not a complete human from a leg.

2. Define the term 'stem cell' in your own words.

3. List the similarities and differences between embryonic and adult stem cells.

4. Explain how the adult cell cloning technique makes it possible to use the patient's own cells to treat a disorder.

5. Explain why there are ethical problems with using embryonic stem cells.

6. Bone marrow transplant is a form of stem cell therapy that has been used since the 1960s to treat patients with leukaemia. Suggest why this kind of stem cell treatment has been available for so long, while other stem cell treatments are still in the development phase.

7. If a couple have a child that is paralysed due to nerve damage, it is possible to mix their sperm and eggs in the lab to create an embryo from which stem cells could be taken and used to cure the child's paralysis. **(a)** What are the advantages to the child of doing this? **(b)** What are the advantages to the parents? **(c)** Should the law allow this kind of treatment? Explain your answer.

8. Stem cells can be taken from the umbilical cord of a baby just after birth. A parent could decide to do this for a baby and have the cells stored in case they are needed later in life to cure a disorder or paralysis. Storage costs money. Evaluate the advantages and disadvantages of this idea.

Genes and alleles

Chromosomes and DNA

Chromosomes are immensely long molecules of DNA (deoxyribonucleic acid) that are found in the nucleus of almost every cell in your body. The DNA molecule is made of millions of subunits of just four types. The order of these subunits forms a code, which can be translated into genetic instructions for building the cell and for maintaining its functions.

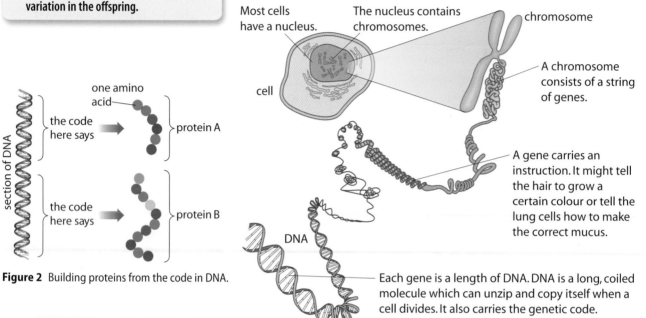

Most cells have a nucleus.

The nucleus contains chromosomes.

chromosome

cell

A chromosome consists of a string of genes.

A gene carries an instruction. It might tell the hair to grow a certain colour or tell the lung cells how to make the correct mucus.

DNA

Each gene is a length of DNA. DNA is a long, coiled molecule which can unzip and copy itself when a cell divides. It also carries the genetic code.

Figure 1 DNA has a **double helix** shape, like a twisted ladder.

section of DNA

one amino acid

the code here says → protein A

the code here says → protein B

Figure 2 Building proteins from the code in DNA.

Genes and proteins

Along a chromosome there are many genes. Each gene is a small section of DNA. A gene codes for a particular sequence of amino acids, and therefore a particular protein. Proteins play many important roles in our bodies.

- Large structures, such as hair and nails, are made from proteins.
- Proteins form the basis for many tissues, such as bone, brain and muscle.
- Within cells, proteins form a large part of structures such as mitochondria.
- Proteins control what can enter and leave the cell.
- Proteins produce the colour of your eyes, hair and skin.
- Enzymes are proteins – they control the rate at which different chemical reactions take place in a cell.

Variation in the code

Variation between individuals occurs because each gene occurs in slightly different forms, called **alleles**. For example, people with blue eyes and people with brown eyes have different alleles of the genes that produce eye colour.

Remember that, as a result of sexual reproduction, you have two sets of chromosomes in your cells, one set from your mother and one set from your father. The *genes* on each chromosome of almost every pair are the same – however

Very few human characteristics are coded for by a single gene. Eye colour is the result of interactions between several genes.

the *allele* for each gene may be different. The characteristic that you have will depend on how those alleles of the gene on each chromosome pair interact.

We can see the variation in the genetic code between individuals not only in their characteristics, but also when we produce an image of their DNA. This is known as **DNA profiling** (also called **DNA fingerprinting**).

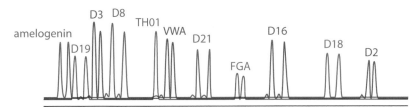

Figure 3 Part of a DNA profile showing a few identified genes. This profile was made using **gas chromatography** (older methods used **gel electrophoresis**).

Only identical twins will have identical DNA profiles, although members of the same family will show similarities. Children have profiles that match half of each parent's profile. Children in the same family may have some or many similarities in their profiles.

Questions

1 A couple have two children: one child has blue eyes, the other has brown eyes. Explain how this is possible.

2 Figure 4 shows three DNA profiles made by gel electrophoresis. The three profiles are from two parents and a child. Identify the child and explain your choice.

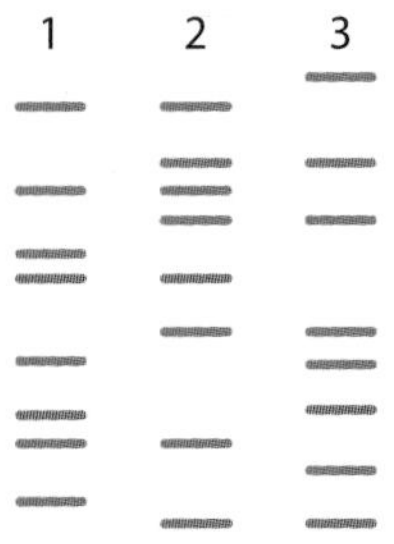

Figure 4 DNA profiles.

3 Gas chromatography is a newer technique for making DNA profiles than gel electrophoresis. What important feature do the techniques have in common?

4 Some people have argued that the police should have a database of DNA profiles from everyone in the UK. Explain an advantage of having a DNA profile database of everyone.

5 Give one social, one economic and one ethical disadvantage of having a DNA profile database for everyone.

6 Do you think police should be allowed to hold DNA profiles for everyone? Explain your answer.

7 Explain fully how genes produce different characteristics, such as dark or fair hair.

8 Explain fully how two children of the same parents could have DNA profiles that are either 100% identical, or 0% identical, and the implications this has for matching DNA profiles with family members in a criminal case.

A*

Science in action

DNA profiling is a key tool in **forensic science**. Tiny samples of tissue such as blood, semen or hair can be used to produce a DNA profile that can help to identify a person who was at the scene of a crime, either by comparing it with a police DNA database or with DNA taken from suspects. However, it cannot specifically identify who carried out the crime.

Practical

DNA can be extracted from any cell with a nucleus, but it is easier to extract from plant cells such as from kiwi fruit.

Science in action

DNA analysis is being used in many areas of science to identify individuals and species, not just in human forensic science. For example, samples taken from whale meat for sale in Japanese markets can show whether or not the meat comes from protected species or from species that are legal to hunt. Comparing the DNA of different species also helps to show evolutionary relationships. For example, evidence from DNA samples taken from fossils of Neanderthal people show that in Europe, where they lived at the same time as *Homo sapiens* (our own species), the two species probably interbred quite frequently.

Inheriting characteristics

Inheriting sex chromosomes

Of the 23 pairs of chromosomes in humans, only one pair differs between males and females. These are the **sex chromosomes** that determine whether you are male or female. In women, both chromosomes in the pair are the same size: they are X chromosomes (see Figure 1 in lesson B2 6.1). In men, one of the sex chromosomes (the Y chromosome) is shorter than the other (an X chromosome).

Gametes contain only one chromosome from each chromosome pair (see lesson B2 6.1), so only one sex chromosome ends up in each gamete. All the eggs that a woman produces will contain an X chromosome, while half the sperm that a man produces will contain an X chromosome and the other half a Y chromosome.

We can explain the inheritance of sex chromosomes using a **genetic diagram** that shows the chromosomes in the body cells, the gametes of the parents, and the full range of possible chromosome combinations in the children.

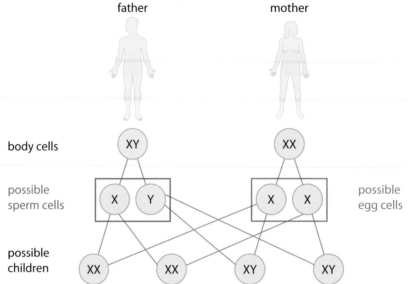

Figure 1 One way of showing the inheritance of sex chromosomes.

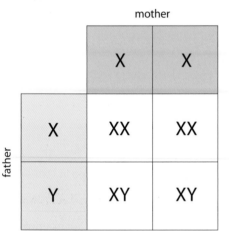

Figure 2 A Punnett square showing the inheritance of sex chromosomes.

Another way of showing the outcome of this uses a different kind of genetic diagram called a **Punnett square**.

Inheriting alleles

We can use Punnett squares like the one in Figure 2 to investigate the inheritance of alleles. In this case, we show the alleles on the chromosome pair for the particular gene we are investigating. Each gamete has only one allele, because the gametes have only single chromosomes rather than pairs. However, the fertilised eggs have two alleles again, one from each parent.

Different pairs of alleles can produce different characteristics. This is because some alleles are **dominant**, where the characteristic they code for always shows whether the individual has one or two copies of the allele. Other alleles are **recessive**, where the characteristic they code for only shows when both chromosomes of the pair have that allele.

Examiner feedback

It is better to draw a Punnett square in an exam rather than a line diagram. This is because the examiner sometimes can't see where the lines are going unless the diagram has been drawn very well.

Examiner feedback

When choosing letters to represent alleles, use something that looks different in capitals and lower case, so it is easy to see the difference. The capital stands for the dominant allele. Always define your use of letters before you answer the question.

In pea plants, the allele for purple flowers is dominant to the allele for white flowers. So when you cross a purple-flowered plant that has two alleles for purple flowers with a white-flowered plant, all the offspring have purple flowers. If you then cross two of the offspring, you can prove that they both contained one allele for purple and one for white flowers because some of the offspring have white flowers. Figure 3 shows that the chance of having purple flowers compared with white is 75%, or a ratio of 3:1.

Mendel's work on inheritance

Gregor Mendel (1822–1884) spent many years studying the inheritance of characteristics in pea plants. He was trying to explain why some characteristics (such as white flower colour) seem to disappear in one generation but reappear in the next. Mendel interpreted his results in terms of separate 'inherited factors', which we now call genes. Although Mendel published his work in 1865, it was largely ignored until about 1900.

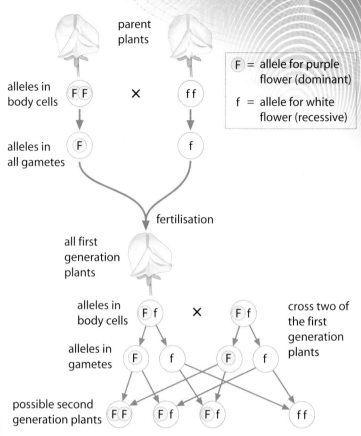

Figure 3 A genetic diagram showing the inheritance of flower colour in pea plants over two generations.

Science skills

Table 1 shows some of Mendel's results. For each cross:

a State what the first generation plants would have looked like.

b Draw a genetic diagram to work out the theoretical result at the second generation.

c Compare the theoretical result from your diagram with the actual result and suggest why there are any differences.

Table 1

Parent varieties	Results at second generation
round seed × wrinkled seed	336 round : 101 wrinkled
tall stem × dwarf stem	787 tall : 277 dwarf

Science in action

Mendel was very careful about how he carried out his experiments. He spent two years creating **pure-breeding** plants (ones that produced only that character when crossed with each other) for each characteristic before he started his crosses. He then transferred pollen from one plant to another using a brush, to make sure he knew which cross had been made. And he repeated each cross many times.

Questions

1 Describe the difference between dominant and recessive alleles.

2 Look at Figure 1. What is the average chance of a couple having a baby boy? Explain your answer.

3 A couple already have three boys. What is the chance that the next child they have is a girl? Explain your answer.

4 Draw a Punnett square for a cross between a purple-flowered plant with two alleles for purple with a white-flowered plant. Describe the proportion of flower colour in the offspring.

5 Draw a Punnett square for a cross between a purple-flowered plant with one allele for purple and one for white crossed with a white-flowered plant. Describe the proportion of flower colour in the offspring.

6 Before Mendel's experiments, many scientists thought that inherited characteristics were blended in the offspring, e.g. a cross between red- and white-flowered plants would give plants with flower colour somewhere between white and red. Explain how Mendel's experiments proved this wrong.

7 Evaluate Mendel's method and explain why it helped him get reliable results.

The causes of extinction

Many frogs and other amphibians, especially tropical species, are threatened with extinction because of a deadly fungal disease.

A series of major volcanic eruptions could produce huge clouds of ash and dust that would block the sunlight and cause a 'global winter'.

Extinction of species

A species evolves as its characteristics change over time. This is driven by **natural selection**, as a result of changes in the physical environment or in the other organisms that live in the same place. If the individuals cannot evolve to survive these changes, the species will become **extinct** (see lesson B1 7.2).

The fossil record shows us that species become extinct, but it cannot tell us why. We need to look for evidence of changes that we think might have caused the extinction. For example, evidence from rocks shows that sea level and temperature have varied greatly over Earth's history. Large or rapid changes in either temperature or sea level could lead to the extinction of species.

The different species in an area affect each other, for example through **predation** or **competition**. About 3 million years ago, North and South America joined together for the first time. Before then the two continents had been separate and many species of plants and animals that lived on them were very different. When they joined, some species moved from one continent to the other.

After this exchange, many species quickly became extinct. Some died out because other species moved into their area and competed successfully for resources of food and space. In other cases new predator species came into an area and killed off whole populations of prey species. New diseases that came with the invaders could also have caused some extinction of local species.

Animals that moved from south to north included ground sloths, terror birds and glyptodonts.

Animals that moved from north to south included camels, cats, wolves, deer, rodents and bears.

Figure 1 The green animals were South American species that spread north, and the blue animals were North American species that spread south when the two continents joined.

Mass extinctions

Sometimes conditions change so rapidly that many species die out at the same time. Such events are known as **mass extinctions**. The worst was about 251 million years ago, when it is thought that over 90% of life on land and in the oceans died out. Another mass extinction about 65 million years ago killed off the last of the dinosaurs.

Most mass extinctions can be linked to rapid changes in sea level or temperature. Some, like the ones 251 million and 65 million years ago, are also linked to volcanic eruptions that lasted over a million years. These would have changed the Earth's climate, affecting both plants and animals. There is some evidence that the extinction 65 million years ago was due to a large asteroid impact.

a The five worst mass extinctions were 65 million, 205 million, 251 million, c.370 million and c.445 million years ago. Look at the graphs of temperature and sea level change in Figure 2 and suggest what part these factors may have played in these extinctions.

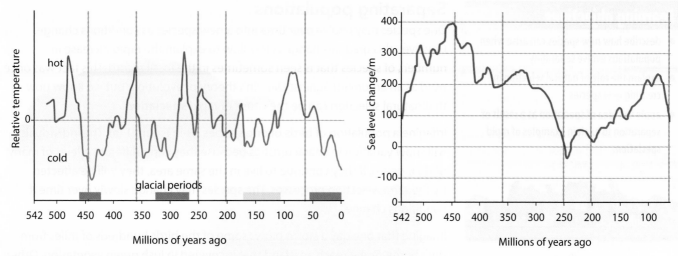

Figure 2 Variation in world temperature and sea level over the past 540 million years.

A changed species

Sometimes the fossil record shows one species disappearing and a similar one taking its place. However, this doesn't necessarily mean that an older species died out and the newer species came from somewhere else. It could be that the original species has evolved so much that it is classified as a new species. Sometimes it is not clear which interpretation is true. This is the case for human evolution (see lesson B2 7.1).

Questions

1 Describe the factors that can cause the extinction of a species.

2 It has been estimated that the volcanic eruptions that occurred about 251 and 65 million years ago both lasted between 1 and 2 million years. Explain how they could cause a mass extinction.

3 In the mass extinction 65 million years ago, about 75% of life on Earth became extinct. Suggest what survived.

4 Woolly mammoths became extinct about 8000 years ago. What evidence would you look for to decide whether the extinction of mammoths was the result of climate change or hunting by humans?

5 When scientists looked at a sample of *Australopithecus* fossil bones, some thought there were two species, one larger than the other, while other scientists thought there was just one species. Suggest why.

6 Before South America was connected to North America, many of the large animals were marsupials. Today, only a few marsupials are found there, while many marsupial species live in Australia. Suggest an explanation for this difference.

7 Much of our understanding about how organisms become extinct has come from looking at the impact of humans on the environment today. Using your answer to question 1, explain how this evidence is limited when we try to apply it to examples of extinction in the fossil record.

8 Explain how a large asteroid impact could cause the extinction of life.

ISA practice: carbohydrase enzymes

Carbohydrase enzymes are used in industry to break down starch into sugar syrup. A manufacturer of sugar syrup has asked some students to investigate the effect of temperature on the time it takes for carbohydrase to break down starch into syrup.

Section 1

1   Write a hypothesis about how you think temperature affects the rate of enzyme action. Use information from your knowledge of rates of reaction to explain why you made this hypothesis. **(3 marks)**

2   Describe how you could carry out an investigation into this factor.

    You should include:

    - the equipment that you would use
    - how you would use the equipment
    - the measurements that you would make
    - a risk assessment
    - how you would make it a fair test.

    You may include a labelled diagram to help you to explain the method.

    In this question you will be assessed on using good English, organising information clearly and using specialist terms where appropriate. **(9 marks)**

3   Design a table that you could use to record all the data you would obtain during the planned investigation. **(2 marks)**

Total for Section 1: 14 marks

Section 2

Two groups of students, Study Groups 1 and 2, investigated the effect of temperature on the breakdown of starch by amylase. Figures 1 and 2 show their results.

Temperature 10°C

Starch present after 0, 2, 4, 6, 8, 10, 12, 14, 16 minutes. No starch after 18 and 20 minutes.

Temperature 20°C

Starch present after 0, 2, 4, 6, 8, 10 and 12 minutes.

No starch 14, 16, 18 and 20 minutes.

Temperature 40°C

Starch present after 0, 2, 4 and 6 minutes.

No starch 8, 10, 12, 14, 16, 18 and 20 minutes.

Temperature 60°C

Starch present after 0, 2, 4, 6, 8, 10 and 14 minutes.

No starch 12, 16, 18 and 20 minutes.

Temperature 80°C

Starch present after 0, 2, 4, 6, 8, 10, 12, 14, 16, 18 and 20 minutes.

Figure 1 Study Group 1's results.

4   (a) Plot a graph of these results. **(4 marks)**

    (b) What conclusion can you draw from the investigation about a link between the temperature and the rate of enzyme action? You should use any pattern that you can see in the results to support your conclusion. **(3 marks)**

    (c) Look at your hypothesis, the answer to question 1. Do the results support your hypothesis? Explain your answer. You should quote some figures from the data in your explanation. **(3 marks)**

    Here are the results of three more studies.

    Figure 2 shows the results from another two students, Study Group 2.

Temperature 10°C

Starch present after 0, 2, 4, 6, 8, 10, 12, 14, 16 minutes. No starch after 18 and 20 minutes.

Temperature 20°C

Starch present after 0, 2, 4, 6 and 8 minutes. No starch 10, 12, 14, 16, 18 and 20 minutes.

Temperature 40°C

Starch present after 0, 2 and 4 minutes. No starch 6, 8, 10, 12, 14, 16, 18 and 20 minutes.

Temperature 60°C

Starch present after 0, 2, 4, 6 and 8 minutes. No starch 10, 12, 14, 16, 18 and 20 minutes.

Temperature 80°C

Starch present after 0, 2, 4, 6, 8, 10, 12, 14, 16 and 18. No starch 20 minutes.

Figure 2 Study Group 2's results.

Figure 3 is a graph drawn from the results of Study Group 3, who carried out another investigation into the effect of a factor on the rate of enzyme action.

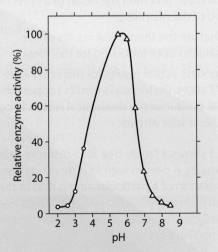

Figure 3 Graph of Study Group 3's results.

Study Group 4 was a group of researchers, who looked on the internet and found Figure 4: a graph showing the effect of temperature on the rate of reaction of a carbohydrase obtained from bacteria.

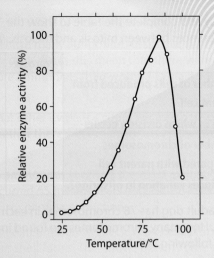

Figure 4 Graph of enzyme activity for a bacterial enzyme.

5 (a) Draw a sketch graph of the results from Study Group 2. *(3 marks)*

(b) Look at the results from Study Groups 2 and 3. Does the data support the conclusion you drew about the investigation in answer to question 5(a)? Give reasons for your answer. *(3 marks)*

(c) The data contain only a limited amount of information. What other information or data would you need in order to be more certain whether the hypothesis is correct or not? Explain the reason for your answer. *(3 marks)*

(d) Look at the results from Study Group 4. Compare them with the results from Study Group 1. Explain how far the data shown supports or does not support your answer to question 5(b). You should use examples from Study Group 4 and Study Group 1. *(3 marks)*

6 (a) Compare the results of Study Group 1 with Study Group 2. Do you think that the results for Study Group 1 are *reproducible*? Explain the reason for your answer. *(3 marks)*

(b) Explain how Study Group 1 could use results from other groups in the class to obtain a more *accurate* answer. *(3 marks)*

7 Applying the results of the investigation to a context.

Suggest how ideas from the original investigation and the other studies could be used by the manufacturers to decide on the best temperature at which to use the carbohydrase enzyme to break down starch, and the best source for this enzyme. *(3 marks)*

Total for Section 2: 31 marks

Total for the ISA: 45 marks

Here are three students' answers to the following question:

Lemmings are mouse-like rodents. The diagram shows the distribution of two species of lemming in North America: *Dicrostonyx torquatus and D. hudsonius.*

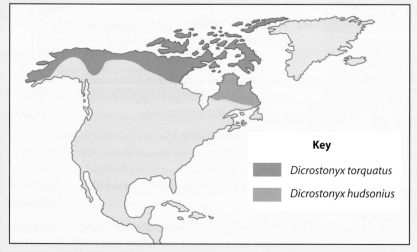

Key

Dicrostonyx torquatus

Dicrostonyx hudsonius

Figure 1 Distribution of Canadian lemmings.

Hudson Bay is a large ocean bay in northern Canada. *D. hudsonius* inhabits the eastern side of the bay and *D. torquatus* inhabits the western side. Before Hudson Bay was formed there was only one species of lemming present in the area.

Explain how the two species of lemming evolved from the original species.

In this question you will be assessed on using good English, organising information clearly and using specialist terms where appropriate. **(6 marks)**

Read the answers together with the examiner comments. Then check what you have learnt and try putting it into practice in any further questions you answer.

B Grade answer

Student 1

> After the bay was formed the lemmings on the east and west sides of the bay did not mix because they were too small to swim across. Because they did not mix they evolved separately. Eventually they could not breed together so they became separate species. Natural selection has occurred.

It would be better to use the correct term, which is 'were geographically isolated'.

Separate evolution is mentioned in the question – no marks are gained for repeating this information.

The candidate has the argument the wrong way round – when they can no longer breed they are separate species.

Examiner comment

The candidate shows some understanding of speciation including an idea of isolation and an idea that different species do not interbreed. But there is very little use of biological terminology. There is no reference to variation. The reference to natural selection needs qualification to gain credit.

Grade answer

Student 2

It would be better to point out that it was the environmental conditions that were different.

It is not clear what 'This' refers to. The reference to natural selection is weak.

The first stage in the formation of two different species is geographical isolation. Here the two groups of lemmings became separated when Hudson Bay was formed. Conditions on the two sides of the bay were different and the two groups adapted to their different surroundings. This led to natural selection. Eventually the two groups became so different that they could not interbreed. They were now two different species.

The candidate seems to be stating that the lemmings intended to adapt to their surroundings.

Examiner comment

A fairly good account that involves most of the main stages, but the account lacks the link between variation and natural selection, which together account for the eventual differences that led to speciation.

Grade answer

Student 3

Good use of appropriate biological terminology.

A good description of variation.

The two groups of lemmings became separated geographically when Hudson Bay formed. Environmental conditions, such as temperature, may have been different on the two sides of the bay. There is natural variation in a lemming population. This is because there is a wide range of alleles in a population. Some variations enabled lemmings to survive in the different conditions. These lemmings would breed and pass their alleles onto the next generation. Survival of the fittest to breed is known as natural selection. In natural selection alleles that enable an organism to survive are selected. Due to natural selection the two populations of lemmings became different. The differences meant that they were unable to interbreed. They had become separate species.

A good description of natural selection.

A good description of how the two populations became different.

Examiner comment

An excellent explanation that gives the main principles underlying speciation: isolation; variation; natural selection; no interbreeding. The principles were described in the correct sequence. Correct biological terminology was used throughout.

Read the whole question carefully.

• Before you begin to answer a question that involves a sequence of events, plan out your answer by writing a key word from each stage in rough. Then sequence the stages before you start your explanation.
• Use the correct biological terminology throughout.
• Make sure you refer to all the information given to you, but do not simply copy information without qualifying it.

Examination-style questions

1   The diagram shows part of a root hair cell.

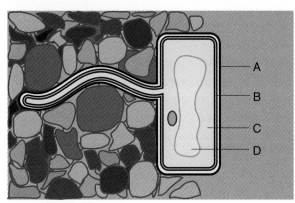

	——— A
	——— B
	——— C
	——— D

(a) Name the structures labelled A, B, C and D.   *(4 marks)*

(b) The diagram below shows four ways in which molecules may move into and out of a cell. The dots show the concentration of molecules.

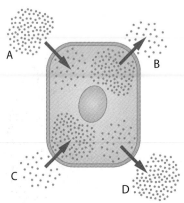

Which arrow – A, B, C or D – represents:

   (i) Movement of oxygen molecules during respiration?

  (ii) Movement of oxygen molecules during photosynthesis?   *(2 marks)*

(c) The digestive system is adapted for the digestion and absorption of food. Describe, in terms of cells, tissues and organs, how the digestive system is adapted for these functions.

   In this question you will be assessed on using good English, organising information clearly and using specialist terms where appropriate.   *(6 marks)*

2   The graph shows the effect of light on the rate of photosynthesis.

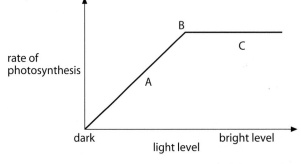

(a) Using the letters A, B and C on the graph, explain the shape of the graph.   *(6 marks)*

(b) Give two ways in which the rate of photosynthesis at C could be increased.   *(2 marks)*

3 Students estimated the number of dandelions on their school's field. They used quadrats, each with an area of 1 m².

The school playing field was rectangular in shape, with dimensions 90 m × 50 m.

The students counted the number of dandelions in 10 quadrats.

(a) Describe *one* way in which the students could have ensured that the quadrats were randomly distributed. *(2 marks)*

(b) The table shows the students' results.

    (i) Calculate the mean number of dandelions per quadrat. *(2 marks)*

    (ii) Use the mean you have calculated in part (i) to estimate the number of dandelions in the field. *(2 marks)*

Quadrat number	Number of dandelions
1	3
2	3
3	6
4	2
5	1
6	2
7	0
8	3
9	2
10	0

(c) The diagram shows the distribution of plants in a lake.

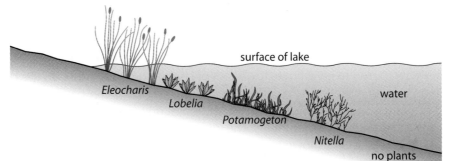

Suggest an explanation for the distribution of:

    (i) *Eleocharis* *(2 marks)*

    (ii) *Nitella.* *(2 marks)*

4 The graph shows the effect of pH on two enzymes from the human digestive system.

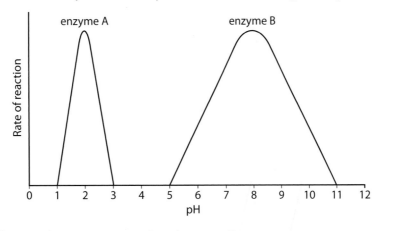

(a) Give two differences between enzyme A and enzyme B. *(2 marks)*

(b) What type of enzyme is enzyme A? Explain the reason for your answer. *(3 marks)*

5 The rate of respiration of an organism can be investigated using a respirometer. A respirometer measures the amount of oxygen used in respiration.

The diagram shows a respirometer containing germinating seeds. The amount of oxygen used during respiration is measured by the movement of the coloured liquid in the capillary tube.

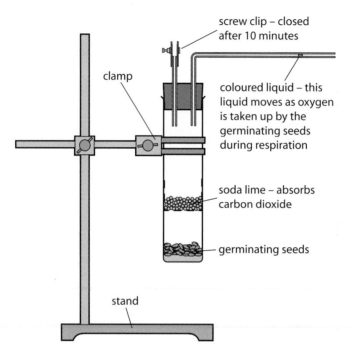

The respirometer was used to investigate the effect of changing temperature on the rate of respiration in germinating seeds. The respirometer was placed in a water bath at 20 °C with the clip open. After 10 minutes, the clip was closed. From this point onwards, the position of the liquid in the capillary tube was recorded every 5 minutes. The investigation was then repeated with the water bath set at 30 °C. The results are shown in the graph.

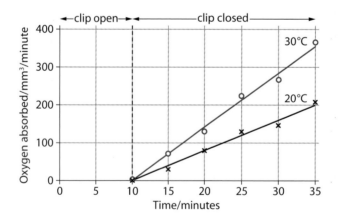

(a) Explain why the respirometer was left in the water bath for 10 minutes before closing the clip. *(2 marks)*

(b) Explain why the liquid moved after the clip was closed. *(2 marks)*

(c) Use the graph to calculate how much oxygen was used per minute at 20 °C. Show your working. *(2 marks)*

(d) Use the results to explain the effect of temperature on the rate of respiration in the germinating seeds. *(2 marks)*

6 The table shows the results of an investigation carried out to find out how breathing changes during exercise.

Activity/step-ups per minute	Mean volume of each breath/cm³	Breathing rate/ breaths per minute	Total amount of air inhaled per minute/cm³
20	500	18	
30	750	25	
40	1100	32	

(a) (i) How many breaths did the person take per minute when exercising at 30 step-ups per minute? *(1 mark)*

(ii) By how much did the volume of each breath increase between 20 and 40 steps per minute? *(1 mark)*

(b) Make a copy of the table. Calculate values for the fourth column and add them to the table. *(3 marks)*

7 The diagram shows the result of one cell division. The parent cell has two pairs of chromosomes.

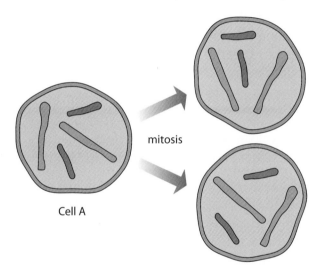

mitosis

Cell A

(a) This cell division is an example of mitosis. Explain why mitosis is important to living organisms. *(2 marks)*

(b) Draw the four cells that would be produced if the parent cell divided by meiosis. *(2 marks)*

8 A mother underwent a new medical procedure as she gave birth. Stem cells were collected from her baby's umbilical cord and stored for possible use in the future. The mother said that using stem cells in this way could save her child's life in the future. She said it was like having an insurance policy for her child.

'Stem cells are undifferentiated cells that act like the master cells of the body,' her doctor said. 'These cells can be used to treat heart problems and various forms of cancer.' Stem cells obtained from embryos produced by IVF can also be used to treat certain diseases.

(a) Explain what is meant by 'stem cells are undifferentiated cells'. *(2 marks)*

(b) Evaluate the use of stem cells to cure human diseases.

In this question you will be assessed on using good English, organising information clearly and using specialist terms where appropriate. *(6 marks)*

Biological systems

Your heart is a remarkable organ. It beats approximately 100 000 times a day, around two and a half thousand million times in your lifetime – and never tires like other muscles. The heart pumps the blood that carries oxygen and nutrients to the body's cells and takes away waste materials. If the heart stops beating as a result of a heart attack, other body organs are also damaged. When the heart is damaged, organs do not get the oxygen and other substances they need. If the damage is extreme surgeons may replace the heart with a healthy one.

Substances are constantly moving in and out of the cells of living organisms. Most substances diffuse in and out of cells. Diffusion is efficient only over very small distances, so large, multicellular organisms need specialised organs for exchanging and absorbing substances. For example, your lungs are specialised for gas exchange. Your breathing system moves air into and out of the lungs, where oxygen diffuses into the blood and carbon dioxide diffuses out of the blood.

Plants have a transport system too. The roots of plants have specialised cells, called root hairs, that are long and thin and have a large surface area for absorbing water and mineral ions. Water and ions are transported from the roots, up the stem to the leaves.

Even though the temperature of the air around you might be freezing in winter and very hot in summer your core body temperature stays at 37 °C. Keeping conditions steady inside the body is called homeostasis. It helps your cells work as efficiently as possible. For example, your body temperature, the amount of glucose in your blood and the amount of water and ions in your body are all kept at steady levels.

Sweating helps to cool your body down, so on a hot day more water is lost as sweat. When you lose water you become thirsty, so you take in more water by drinking more fluids. When you sweat you lose ions (salts) as well as water. Sports drinks help to replace both the water and the salts.

Test yourself

1   How does the oxygen needed for respiration reach cells in your body?
2   Which structure controls the entry and exit of substances in cells?
3   Why do particles in solution move?
4   What is the role of plant roots in photosynthesis?
5   Give one reason why it is important for conditions inside the body to be kept constant.

Objectives

By the end of this unit you should be able to:

- explain how materials get into and out of animals and plants
- explain how materials are transported around large organisms
- explain how our bodies keep internal conditions constant
- evaluate the development and use of artificial aids to breathing including the use of artificial ventilators
- evaluate the claims of manufacturers about sports drinks
- evaluate the optimum conditions for transpiration in plants
- evaluate data on the production and use of artificial blood products
- evaluate the use of artificial hearts and heart valves
- evaluate the use of stents
- evaluate the advantages and disadvantages of treating kidney failure by dialysis or kidney transplant
- evaluate modern methods of treating diabetes.

Diffusion and osmosis

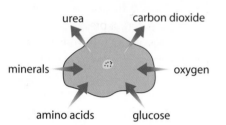

Figure 1 Substances entering and leaving a cell.

Movement in and out of cells

Many different types of substance move in and out of the cells of living organisms. Figure 1 shows some of the substances that regularly pass in and out of cells. All cells are surrounded by a cell membrane. It forms a barrier that any substance entering or leaving a cell must pass through. There are three main ways in which substances move in and out of cells: **diffusion**, **osmosis** and **active transport**.

Molecules of gases, such as oxygen and carbon dioxide, are moving about all the time. The molecules of a solution, such as glucose dissolved in water, are also moving in all directions. When molecules move, they spread themselves out evenly. This causes them to move from a higher concentration to a lower concentration until the concentrations become the same. This process is called diffusion. As they move from a higher to a lower concentration, molecules diffuse down a **concentration gradient**. Most substances move into and out of cells by diffusion.

Osmosis

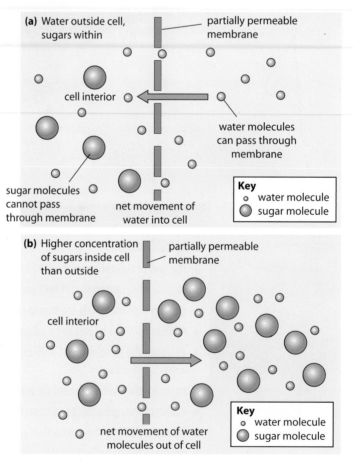

Figure 2 Osmosis.

Water molecules can diffuse through partially permeable membranes, but sugar molecules and most ions generally cannot. A cell membrane has specialised molecules to enable some ions and molecules to move into the cell.

A dilute solution contains more water molecules than a concentrated solution. If a partially permeable membrane separates two solutions of different concentrations, there will be net movement of water molecules from the dilute solution to the concentrated solution. Diffusion of water molecules across a partially permeable membrane is called osmosis.

Osmosis explains the movement of water molecules across plant and animal cell membranes. In a plant, water moves from the soil into root-hair cells, because soil water is a more dilute solution than the solution in the cytoplasm of the cell.

Osmosis also happens between animal cells. However, animal cells have no strong cell wall to protect them, so if too much water enters them they burst. If animal or plant cells lose too much water, the cytoplasm shrinks and the cells cannot function properly.

 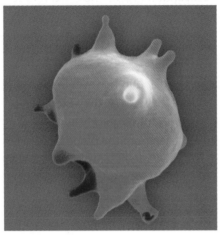

The blood cell in the left-hand photo was in a dilute solution. The right-hand photo shows a blood cell that has changed because it was in a concentrated solution.

Questions

1. What is meant by a 'partially permeable membrane'?
2. What is meant by 'concentration gradient'?
3. Explain how cell membranes control the particles that pass through them.
4. A person drinks a large volume of water very quickly. What effect will this have on the red cells in the blood? Explain why.
5. If we place animal cells in pure water they burst. But plant cells do not burst. Explain why.
6. If a person loses a lot of blood they are given a transfusion of a solution containing ions. What concentration should this solution of ions be? Give the reason for your answer.
7. To prepare fruit salad, a cook cuts up different types of fruit and sprinkles sugar over the pieces. After two hours the fruit is surrounded by syrup (concentrated sugar solution). Explain why.
8. Water moves from the soil into a root hair cell and then through other cells towards the centre of a root. Explain this in terms of osmosis.

Examiner feedback

Remember that the cell wall is permeable to water, ions and sugars. It therefore takes no part in osmosis.

Taking it further

As water continues to enter plant cells by osmosis, the cell contents begin to exert a pressure on the cell wall. This is known as **turgor** pressure. In young plants turgor pressure is the main means of support. When plant cells are placed in a concentrated solution, water leaves the cytoplasm and vacuole by osmosis. Eventually the cytoplasm begins to shrink away from the cell wall. This is known as **plasmolysis**.

Science in action

The main problem facing people on lifeboats in the open ocean is dehydration, even though they are surrounded by water. This is because the concentration of **ions** in seawater is about four times greater than that in our body fluids. Drinking one litre of seawater causes the concentration of ions in our body fluids to rise by about 10 per cent. The effect of the rise is to cause water to move out of the body cells by osmosis, making the cells shrink.

Many lifeboats are now equipped with a reverse osmosis machine to provide drinking water from salt water. Figure 3 shows how it works.

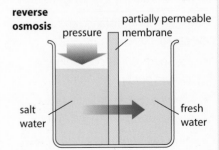

Figure 3 A reverse osmosis machine.

Sports drinks and active transport

Learning objectives

- explain why some sports drinks improve performance
- evaluate the claims of manufacturers about sports drinks
- explain the process of active transport.

Science in action

Water intoxication (hyponatremia), is caused by drinking excessive amounts of plain water, which causes a low concentration of sodium in the blood. Research has found that, in long-duration endurance events such as ultra-marathons, many competitors finish with low blood sodium concentrations. Those at most risk are those who are on the course the longest, because they tend to drink the most water during the event. Runners who drink extra fluids in the days before the race or those who stop at water stops during the race are also at increased risk of hyponatremia. Investigations have also shown that about 15% of marathon finishers develop hyponatremia from drinking too much water.

Athletes in long-distance endurance events are advised to:

- use sports drinks containing sodium during the event
- increase their daily salt intake several days prior to the event
- try not to drink more than they sweat: about 1 cup of fluid every 20 minutes.

Rehydration

When we sweat we lose a lot of water, but not quite as many ions. The concentration of ions in sweat is about 1–2 g per litre, which is less than in blood. This leaves us with more ions in our blood than normal. If the balance of ions and water changes in our bodies, cells do not work as efficiently.

Any type of drink will help you to **rehydrate** (replace the water you have lost).

Most soft drinks consist mainly of water, with sugar and flavourings added. Some also contain ions, but the concentration of sugars and ions varies enormously between soft drinks. The sugar content can vary between zero and over 100 g per can, and sodium ions between zero and 150 mg per can. It is important that drinks contain sodium ions since sodium ions are essential for the healthy functioning of most of the body's cells.

Athletes in endurance events such as the marathon need to replace the water and ions they lose as they exercise. Sports drinks help athletes to replace both water and ions as quickly as possible. Athletes also need to replace the glucose that is taken from the blood by the muscles during exercise. The glucose is used to release energy via respiration. Most athletes will drink sports drinks during an event. All sports drinks contain water and ions. Most sports drinks also contain sugar (usually glucose).

Research has shown that the most effective sports drinks contain between 6% and 8% sugar and 120 mg/l sodium ions. Drinks containing these concentrations of sugar and sodium ions are absorbed more quickly than other drinks.

Science skills

Sports scientists investigated the effect of different drinks on the performance of cyclists. The cyclists were given one of four different types of drink and asked to cycle 8 km as fast as possible. The graph shows the results of the investigation.

a Suggest the composition of the placebo.

b i Which was the most effective drink?

　ii Suggest why this was the most effective drink.

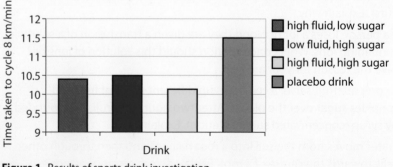

Figure 1 Results of sports drink investigation.

Active transport

Sometimes the body cells need to move substances from an area of lower concentration to an area of higher concentration. This means moving them against a concentration gradient. This requires energy, which is supplied by respiration. The energy is used to move substances through special channels in the cell membrane. Moving molecules using energy from respiration is called **active transport**. Respiration is carried out in the mitochondria, so cells that use energy for active transport have lots of mitochondria.

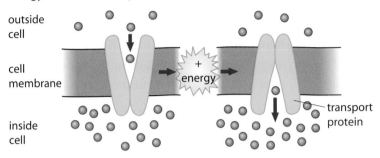

Figure 2 Special proteins create channels for active transport in the cell membrane.

Plants need nitrate ions to make proteins for growth. There is a very low concentration of nitrate ions in soil water surrounding the root. Plants use energy to actively transport nitrate ions across the cell membrane into root cells, against a concentration gradient.

Examiner feedback

Questions on active transport often test your understanding of mitochondria and respiration. If a cell has greater than normal numbers of mitochondria, it has a greater rate of respiration and releases greater amounts of energy. So, cells whose main function is actively transporting substances will have greater numbers of mitochondria.

Sometimes you will be given data involving the use of respiratory poisons. Remember that if respiration stops, active transport will also stop.

Why do you lose a lot of sweat during a good workout?

Science in action

Posters for a sports drink say that it is the 'water designed for exercise'. The eye-catching television advert for this drink shows an athlete made of water, running, doing cartwheels and back flips, diving into a large pool and swimming away. A voice says: 'Imagine water redesigned for exercise and for better hydration than water alone'. The drink contains 2 g carbohydrate and 35 mg sodium per 100 ml, and provides 10 calories. The ingredients are: water, glucose syrup, citric acid, acidity regulators, flavouring, sweeteners and vitamins.

c Are the makers of this sports drink justified in calling it 'water redesigned for exercise'? Explain the reasons for your answer.

Questions

1 What are the main constituents of soft drinks?

2 What are the main differences between a soft drink and a sports drink?

3 Explain why losing a lot of sweat can affect an athlete's performance.

4 Explain why drinking a sports drink can improve an athlete's performance.

5 Give two differences between osmosis and active transport.

6 **(a)** Explain what is meant by 'against their concentration gradient'. **(b)** Why can't this movement be accomplished by diffusion?

7 **(a)** Which organelles in a cell carry out the reactions of respiration? **(b)** Suggest how you would identify a cell that carries out a lot of active transport.

8 Some manufacturers produce drinks containing 'super-oxygenated water'. The manufacturers claim that the extra oxygen in these drinks improves performance by up to 35%. Scientists investigated what happened when runners thought they were getting a performance boost in the form of super-oxygenated water. The athletes completed three separately run 5 km time trials, with half the group drinking a large glass of plain bottled water and the others taking what they thought was super-oxygenated water (but was, in fact, tap water) before they started. Results showed that runners covered the distance an average of 83 seconds faster when they thought they were drinking super-oxygenated water.

Evaluate the report of the above investigation.

Exchanges in humans

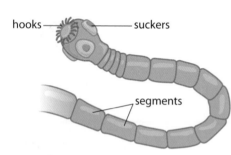

Figure 1 Tapeworms live in intestines and absorb the soluble foods that surround them.

Size and exchanges

Living organisms must exchange substances with the environment. Diffusion is only efficient over short distances, so larger organisms have evolved other mechanisms for exchanging substances.

Small multicellular organisms have adapted by increasing their surface area and decreasing the distance substances have to diffuse. For example, tapeworms live in the intestines of animals. Their bodies are composed of hundreds of flat segments. Being long and flat gives the tapeworm a very large surface area. This allows them to absorb soluble food directly from their surroundings. Being flat means that the distance food has to diffuse to reach the centre of each segment is very short. At no point in a tapeworm's body does food have to diffuse further than 0.1 mm.

In larger organisms diffusion cannot supply the cells at the centre of the body with food and oxygen. Humans and other mammals have overcome this problem by evolving specialised organs with internal surfaces, such as **alveoli** in the lungs and **villi** in the intestines. Animals with internal exchange surfaces usually have a blood system to transport materials between the exchange surfaces and the tissues.

Alveoli

As you breathe you take in the oxygen you need for respiration, and you also get rid of waste carbon dioxide. In your lungs you exchange the oxygen you need for the carbon dioxide you don't need. This is called gas exchange.

Oxygen and carbon dioxide diffuse rapidly between the air in your lungs and your blood. This is because your lungs are highly specialised for exchanging gases. To carry out gas exchange efficiently, the surface of your lungs needs to have the following features.

- Thin walls so that gases diffuse across only a short distance.
- A good blood supply to transport oxygen and carbon dioxide to and from body tissues.
- A large surface area for diffusion.

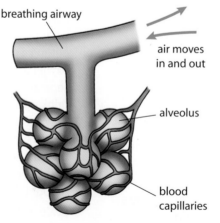

Figure 2 Airway, alveoli and blood supply.

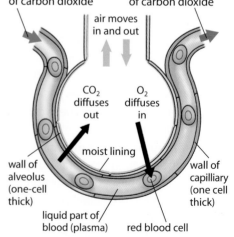

Figure 3 Gas exchange in a single alveolus.

Figure 2 shows the relationship between aveoli and the blood system. Your breathing system takes air in and out of your body. This provides a regular supply of air containing oxygen, and removes air containing carbon dioxide.

The airways of the breathing system end in very small air sacs called alveoli. The walls of the alveoli are where gas exchange happens, and they provide an extremely large surface area for gas exchange. The total surface area of all the alveoli is about 80 m². You can see from Figure 3 that the alveolar walls are only one cell thick and each alveolus is surrounded by blood capillaries. The air in each alveolus is very close to the blood flowing in capillaries. This means that oxygen has to diffuse only a short distance to move from the alveolus into the bloodstream, and carbon dioxide has to diffuse only a short distance in the opposite direction.

Villi

The wall of your small intestine is very efficient at absorbing food. It is a specialised tissue for absorption. The small intestine can absorb food efficiently for the following reasons.

- Its inner surface contains many tiny folds called villi. The large number of villi produces a very large surface area for absorption.

- Each villus contains many blood capillaries to transport absorbed food from the small intestine to the rest of the body.

- Each villus is very thin, so that food molecules diffuse over only a short distance to reach the bloodstream. Figure 4 shows how these features allow absorption to take place efficiently.

Soluble foods molecules are absorbed into the outer cells of the villi by both diffusion and active transport (see lesson B3 1.2). The soluble food molecules are then moved to the blood capillaries. The blood system distributes the soluble food molecules to the rest of the body.

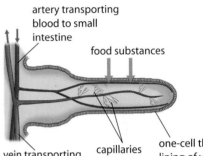

Figure 4 How a villus absorbs food. Each cell of the villus wall has microvilli that further increase the surface area for absorption.

Questions

1 Explain the effect of having millions of alveoli in the lungs.

2 What features of the alveoli provide a short diffusion pathway?

3 Food molecules diffuse over a short distance to reach the blood. What features of the small intestine provide the short diffusion distance?

4 **(a)** Suggest how steep concentration gradients for oxygen and carbon dioxide are maintained in the lungs. **(b)** Explain why this is important for effective gas exchange in the lungs.

5 What features of the small intestine maintain a concentration gradient of food molecules?

6 Describe one feature of the villus cell that indicates that some food molecules are absorbed by active transport.

7 Explain the advantage of having lining cells with a highly folded cell membrane.

8 Explain fully how large animals have evolved to cope with the problem of exchanging materials with the environment.

Include information on:
- how the surface area available for absorption is increased transport mechanisms
- removal of waste substances from tissues.

Gaseous exchange in humans

Learning objectives

- describe the structure of the human breathing organs
- explain how the action of muscles causes air to enter and leave the lungs
- evaluate the development and use of artificial aids to breathing, including the use of artificial ventilators.

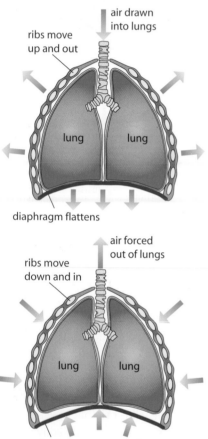

Figure 1 Breathing in (top) and out (bottom).

Examiner feedback

Do not confuse respiration and breathing. Breathing is the movement of air into and out of the lungs. Respiration is the series of reactions that occur inside cells to release energy from sugars.

Breathing

Your lungs are in the upper part of your body, your **thorax**, protected by the **ribcage**. Below the lungs is a sheet of fibre and muscle called the **diaphragm** that separates the thorax from the **abdomen** below. Movements of the ribcage and diaphragm cause you to breathe in and out. The movement of air in and out of the lungs is known as **ventilation**.

To breathe in, two sets of muscles contract at the same time.

- The **intercostal muscles** (muscles between the ribs) contract, pulling the ribs upwards and outwards.

- The diaphragm muscles contract, pulling the central part of the diaphragm downwards.

These changes in the position of the ribs and diaphragm increase the volume of the thorax and expand the lungs. Increasing the volume of a gas decreases its pressure, so the air inside the lungs is now at a lower pressure than the air outside the body. The difference in pressure causes air to move from the outside into the lungs. The process of breathing in is known as **inspiration**.

To breathe out, both the diaphragm and the intercostal muscles relax. Elastic recoil of the lungs and the thorax wall return the lungs to their original size. This decreases the volume of air in the lungs. The pressure of the air inside the lungs is now greater than that of the air outside, so air moves out of the lungs. This process is known as **expiration**.

Some of the air in your lungs is replaced each time you breathe. This keeps a relatively high concentration of oxygen and a relatively low concentration of carbon dioxide in the lungs. At the same time the blood in the capillaries is continually circulating, bringing blood to the lungs with a high concentration of carbon dioxide and a low concentration of oxygen.

Artificial aids to breathing.

A healthy person breathes automatically twenty-four hours each day. However, spontaneous breathing may stop due to disease or injury. If this happens, the patient can be helped to breathe using a mechanical device. There are two main types of device: a machine called a ventilator and a bag that can be compressed manually.

There are two types of ventilator: **negative-pressure ventilators** which cause air to be 'sucked' into the lungs and **positive-pressure ventilators** which force air into the lungs.

Negative-pressure ventilators

The first type of negative-pressure ventilator was the **iron lung**. This type of ventilator was first produced during the 1920s. It was developed for widespread use during epidemics of **poliomyelitis** in the 1940s. In some cases of poliomyelitis, the nerves supplying the breathing muscles ceased to function and the patient stopped breathing. Iron lungs kept many of these patients alive, often for long periods, until the patients recovered.

An iron lung is essentially a large tank enclosing the whole of a patient's body except for the head and the neck. A rubber seal around the patient's neck keeps the tank airtight. A pump removes air from the tank, creating a vacuum. This causes the patient's thorax to expand. Pressure in the lungs decreases, and air from outside moves into the lungs. The vacuum is then released and elastic recoil of the lungs and thorax forces air out of the lungs.

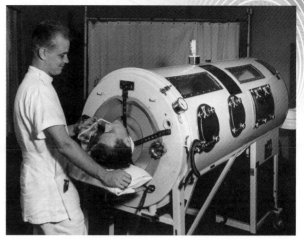
An iron lung in use in the 1920s.

Positive-pressure ventilators

Positive-pressure ventilators were first developed during the 1950s. These machines force air into a patient's lungs through a metal tube inserted through the mouth into the windpipe. The ventilator is most commonly used in operations, during which a patient's muscles are deliberately made to relax to make surgery easier. It is also used to sustain breathing of patients in intensive care units. For long-term use the tube is inserted surgically through the neck into the trachea.

Hand-controlled ventilators

The most common type of hand-controlled ventilator is the bag-mouth-mask ventilator. These ventilators are most often used by paramedics as **resuscitators** to help patients who have stopped breathing after accidents or drug overdoses. Air is supplied via a bag, which can be squeezed manually or operated by a pump.

A modern ventilator.

Questions

1. What is meant by ventilation?
2. Name the muscles that bring about ventilation.
3. Explain the mechanism of inspiration in terms of pressure and volume changes.
4. Explain why air moves out of the lungs during expiration.
5. Explain why some patients who get poliomyelitis need artificial ventilation.
6. Explain how paramedics resuscitate a person who has stopped breathing.
7. Describe how the treatment of breathing failure has changed over the years.
8. Explain how an iron lung keeps a patient alive.
9. Compare and contrast the iron lung and the positive-pressure ventilator machine for long-term care of patients.

A*

A bag-mouth-mask ventilator or BVM is a normal part of the rescuscitation kit for an ambulance crew.

Exchange systems in plants

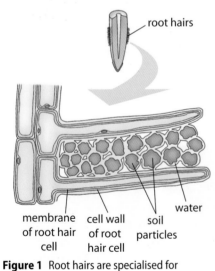

Photomicrograph of the underside of a leaf, showing the stomata (pink).

Water and ions

Green plants make their own food by photosynthesis (see lesson B2 2.1). One of the vital ingredients for photosynthesis is water. As well as being used in photosynthesis, large amounts of water are used to transport materials around the plant.

Water and mineral ions are absorbed from the soil by the roots of plants. The parts of the root that are specialised for absorption are the root hairs. Root hairs are found just behind the growing tip of the roots. Each root hair is a tube-like extension of a cell. By growing between soil particles, each root hair is surrounded by water containing dissolved ions. This means that water and ions diffuse over only a very small distance to reach the root hair cell. By having lots of root hairs, the surface area for absorption of water and ions is greatly increased.

Movement in and out of leaves

Leaves can carry out photosynthesis efficiently because they are adapted to absorb large amounts of carbon dioxide and sunlight.

Leaves are highly flattened, which gives them a large surface area. This allows them to absorb carbon dioxide very efficiently. The surface area is increased further by many internal air spaces. Figure 2 shows the path that carbon dioxide takes. Molecules of carbon dioxide diffuse into the leaf through tiny pores called **stomata**. It then diffuses through the air spaces within the leaf to reach photosynthesising cells.

Transpiration

Plants lose water from the surface of their leaves continuously. This is called **transpiration** (see Figure 3). Most of the water vapour lost by transpiration is through stomata.

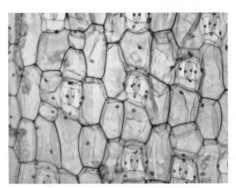

Figure 1 Root hairs are specialised for absorbing water and ions.

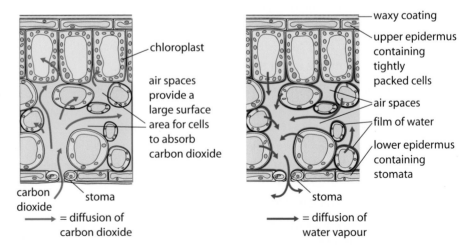

Figure 2 Diffusion of carbon dioxide into a leaf. **Figure 3** Diffusion of water out of a leaf.

Cutting down water loss

The amount of water lost by transpiration increases in hot, dry and windy conditions. As water is lost from the leaves, more water is absorbed into the plant at its roots. When transpiration rates increase, the roots may not be able to absorb enough water from the soil to replace the water lost at the leaf surface. When this happens the plant wilts. To prevent wilting, plants can close their stomata to reduce the rate of transpiration.

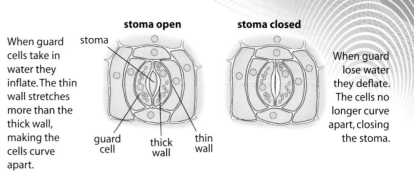

stoma open   **stoma closed**

When guard cells take in water they inflate. The thin wall stretches more than the thick wall, making the cells curve apart.

stoma

guard cell   thick wall   thin wall

When guard lose water they deflate. The cells no longer curve apart, closing the stoma.

Figure 4 Each stoma is surrounded by a pair of sausage-shaped guard cells. These control water loss from the leaf.

Science skills

A potometer was used by a group of students to measure the rate of transpiration in different conditions. The same plant was used throughout the investigation. The conditions were altered by using a hair drier to blow hot or cold air over a leafy shoot. The air bubble was positioned at the start of the scale during each recording. The results are shown in Table 1.

Table 1 Results of student transpiration investigation.

Time/ min	Total length of roots produced/cm		
	Still air	Air at room temperature being blown by hair dryer	Warm air being blown by hair dryer
0	0	0	0
5	1	6	10
10	3	11	20
15	4	16	27
20	5	21	33
25	8	24	38
30	10	27	42

a Describe the pattern of results for the plant at room temperature being blown by the hair drier. Suggest an explanation or this pattern.

b Under which conditions is the rate of transpiration the highest?

c How could the students improve the reliability of their investigation?

Science skills

leafy shoot

syringe

bubble at start of scale

scale

5 4 3 2 1

vaseline

capillary tube of known diameter

water

Figure 5 Measuring transpiration.

Figure 5 shows an apparatus called a **potometer**. This measures the rate at which a plant takes up water. As the plant loses water by transpiration, more water is taken up, making the air bubble move along the capillary tube. The rates of transpiration in different conditions can be compared by measuring how fast the air bubble moves.

Questions

1 What effect do root hair cells have on the surface area for absorption? Explain why this is this important for the plant.

2 Dissolved ions in soil water are at a lower concentration than the ions in the root cell cytoplasm. Suggest what method plants use to absorb mineral ions. Explain the reason for your answer.

3 Effective exchange surfaces have a large surface area. Explain how this is achieved in a leaf.

4 What features of the upper surfaces of leaves help to reduce the loss of water by transpiration?

5 Suggest why stomata are found only on the lower surface of most leaves.

6 **(a)** Name three gases that pass through the stomata. **(b)** In which direction does each of these gases usually move at noon on a hot, sunny day? Give the reasons for your answers.

7 Suggest conditions that would cause a plant to close its stomata. What effect would this have on the rate of photosynthesis? Explain your answer.

8 Describe the factors that affect the rate of transpiration. Explain why each factor has its effect.

A*

The heart and circulation

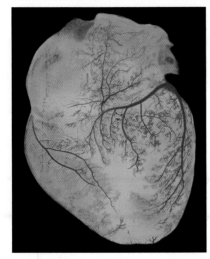

This view of the heart shows the blood vessels carrying blood to the cardiac muscles.

Pumping blood

Blood flows around your body through a series of blood vessels. These make up your **circulatory system**. Blood is kept flowing around the system by the pumping action of your heart. Your heart beats because the muscles in its walls contract and then relax. When the muscular walls contract, blood is forced out of the heart under high pressure. When the muscular walls relax, the heart fills up with more blood, ready to be pumped out again. Beating at between 60 and 80 beats per minute at rest, your heart is working all the time without pause.

The heart pumps blood out through wide blood vessels called **arteries**. The arteries branch again and again to give narrower vessels that spread throughout the body. Within the organs, the smallest blood vessels are only the width of a red blood cell. These vessels are called **capillaries**.

As they leave the organs, the capillaries join together again and again to form wider vessels called **veins**. By the time they reach the heart all the veins have joined up to form only two large blood vessels.

The heart

The heart is divided into two sides, right and left. Each side consists of two chambers: an upper **atrium** and a lower **ventricle**. Each side has a **valve** that allows blood to flow from the atrium into the ventricle, but prevents backflow. Blood is returned to the heart by two **veins**: the **vena cava** and the **pulmonary vein**. Blood is pumped out of the heart through two **arteries**: the **aorta** and the **pulmonary artery**. Valves at the base of the pulmonary artery and the aorta prevent the backflow of blood into the heart.

Figure 1 shows how the heart works as a pump.

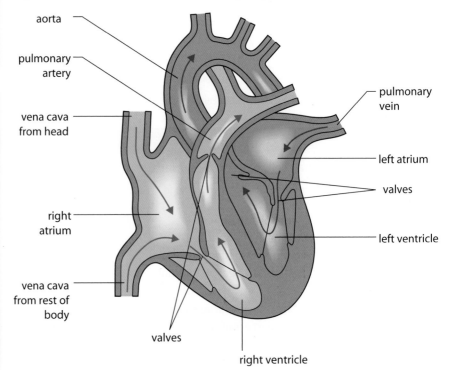

Figure 1 The pumping action of the heart.

Double circulation

The numbers on Figure 2 show the path of blood as it flows through the heart and around the body.

1. Blood from the right ventricle of the heart is pumped out through the pulmonary artery to the lungs.

2. In the lung tissue, oxygen diffuses into the blood and carbon dioxide diffuses out. The blood has become oxygenated.

3. Oxygenated blood from the lungs then returns via the pulmonary vein to the left atrium.

4. Oxygenated blood is pumped through the aorta and arteries to the rest of the body.

5. In respiring tissues, oxygen diffuses from the blood into body cells, and carbon dioxide diffuses from body cells into the blood. The blood has become deoxygenated.

6. Deoxygenated blood is returned by the vena cava to the right atrium.

This is called a double circulatory system because blood travels through the heart twice as it flows around the body. By having a double circulation, oxygenated blood is separated from deoxygenated blood.

Figure 2 The double circulatory system.

Science skills

Figure 3 shows changes in the pressure of the blood as it passes around the body.

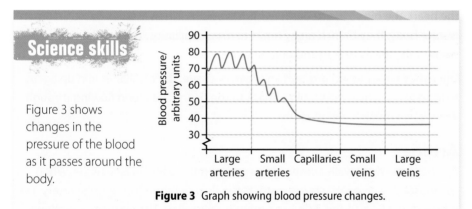

Figure 3 Graph showing blood pressure changes.

a What is the range of blood pressure in the largest arteries?

b In which type of blood vessels does the blood pressure fall the most?

c What is the pressure of blood as it flows back into the heart?

Questions

1 What type of tissue makes up most of the heart?

2 Name the two types of chamber in the heart.

3 Define the following. **(a)** Artery. **(b)** Vein. **(c)** Capillary.

4 Explain why the heart contains valves.

5 Describe the changes in the composition of the blood as it flows through the lungs.

6 Explain what is meant by a double circulatory system.

7 Explain how the structure of the heart is adapted to its function.

Blood vessels

Learning objectives

- relate the structure of arteries, veins and capillaries to their functions
- explain how substances are exchanged between the blood and the tissues
- interpret data relating to structural features of blood vessels

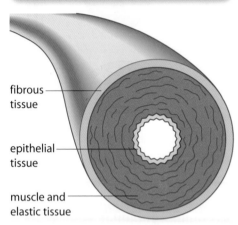

fibrous tissue

epithelial tissue

muscle and elastic tissue

Figure 2 The structure of an artery.

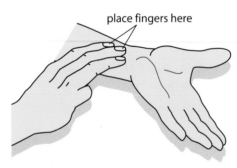

place fingers here

Figure 3 You can feel an artery stretching and recoiling as blood flows along an artery. This is your pulse.

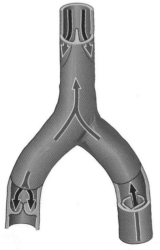

Figure 4 Valves in veins.

The circulatory system

Blood is pumped around your body through a system of blood vessels. There are three types of blood vessel: arteries, veins and capillaries.

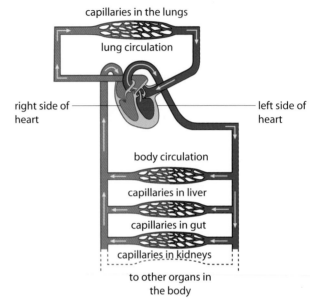

capillaries in the lungs

lung circulation

right side of heart

left side of heart

body circulation

capillaries in liver

capillaries in gut

capillaries in kidneys

to other organs in the body

Figure 1 Plan of the blood circulation in humans.

Blood flows out of the heart through arteries. Because it has been pumped from the heart, the blood in arteries is at high pressure. When arteries get to an organ in your body, they branch many times to form capillaries. The capillaries are where substances pass in and out of the blood.

Once they have passed through an organ or tissue, the capillaries join up again to form veins, which return blood back to the heart. The blood flowing through veins is at a much lower pressure.

Arteries

Arteries have thick walls composed largely of muscle and elastic tissue. When the muscle tissue contracts it constricts the flow of blood through the artery. This is how blood flow to the different organs is controlled. The elastic tissue allows the artery to expand when blood is forced into it from the ventricles. When the ventricles relax, recoil of the elastic tissue keeps the pressure of the blood high.

Taking your pulse

Blood does not flow smoothly through arteries. Every time the heart muscles contract, a surge of blood passes along the arteries, causing the artery walls to bulge slightly. When this happens the walls of arteries become stretched. The walls then spring back (recoil) as the heart muscles relax. You can feel arteries stretch and recoil when you feel the pulse in your wrist.

Veins

Veins have much thinner walls than arteries. The blood flowing through them is at a much lower pressure than that flowing through arteries. To prevent backflow in long veins such as those in the legs, they have valves.

Capillaries

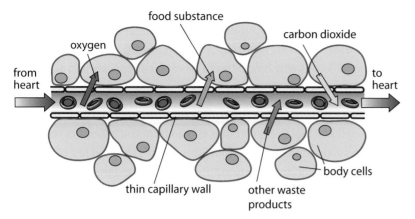

Figure 5 Capillaries exchange substances with tissues.

The smallest blood vessels are called capillaries. This is where substances pass in and out of the blood. Capillaries are so narrow that only one red blood cell at a time can be squeezed through. They also have very thin walls, just one cell thick, that allow substances to pass in and out easily. Substances needed by body tissues, such as soluble foods and oxygen, pass out of the blood, through the capillary wall and into cells. Substances produced by cells, such as carbon dioxide and other waste substances (see lesson B3 3.2), pass into the blood through the walls of the capillaries.

Science skills

Table 1 gives data about the blood vessels in a small mammal.

Table 1 The blood vessels of a small mammal.

Blood vessels	Mean diameter/cm	Mean length/cm	Total cross-sectional area/cm²	Total volume/cm³
main arteries	0.1	10.0	5	50
small arteries	0.002	0.2	125	25
capillaries	0.0008	0.1	600	60
main veins	0.24	10.0	27	270
other blood vessels				525

a How many times wider than a capillary is a main artery?

b What percentage of the blood is in the capillaries at any one time?

c The main arteries and the main veins have the same mean length. Use data from the table to explain why the volume of blood in the main veins is much higher than that in the main arteries.

d Use information from the table to explain why the pressure of blood in the capillaries is much less than that in the main arteries.

Questions

1 Give the function of each of the following in the wall of an artery. **(a)** Muscle tissue. **(b)** Elastic tissue.

2 Give two differences between the structure of an artery and that of a vein.

3 Describe the structure of a capillary.

4 List the exchanges that occur between the blood in a capillary and the tissues.

5 Explain why cells exchange substances with capillaries rather than with arteries or veins.

6 Explain why veins in the legs have valves.

7 Describe how a molecule of oxygen gets from the air to a muscle cell.

The blood

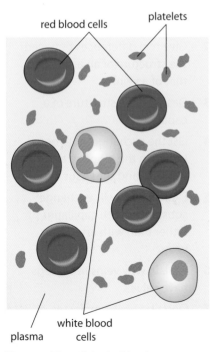

Blood in a bag: what might this be used for?

Your blood

Your blood provides all the cells of your body with the materials they need, as well as removing waste materials. Your cells would soon stop working without a good supply of blood. This is why blood has to be replaced quickly if someone loses a lot of blood in an accident.

Your blood looks like a red liquid, but the liquid part is not red – it is a straw-coloured liquid called plasma. The red colour comes from the billions of red blood cells that are suspended in the plasma. Both the plasma and red blood cells play an important role in transporting substances around the body.

The plasma also carries other cells. **White blood cells** form part of the body's defence system against pathogens. They act in three main ways: producing antibodies, producing antitoxins and engulfing pathogens. **Platelets** are small fragments of cells. They have no nucleus. Platelets help to reduce blood loss by producing a clot at the site of a wound.

Transporting food and waste materials

Plasma is the liquid part of blood. Plasma contains mainly water with a number of dissolved substances, including the following.

- Soluble sugars, amino acids, fatty acids and glycerol are products of digestion in the small intestine. These soluble products are absorbed from the small intestine and transported in plasma to other body organs.

- Carbon dioxide is produced as a waste product of respiration. Carbon dioxide is transported in plasma from respiring cells to the alveoli in the lungs. As blood flows around the alveoli, carbon dioxide is removed from the blood by diffusion and breathed out.

- Urea is a waste product formed from excess amino acids. Urea is made in the liver and transported in blood plasma to the kidneys, where it is removed from the body in urine.

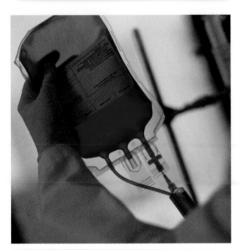

Figure 1 The cells in the blood.

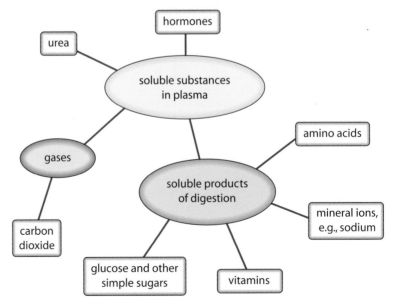

Figure 2 Some of the substances that dissolve in blood plasma.

Transporting oxygen

Your blood contains an enormous number of red blood cells. These are very specialised cells that transport oxygen to all the cells of your body. Red blood cells are shaped like a biconvex disc. They are so specialised that they have no nucleus. Each cell is packed with a protein called **haemoglobin**. It is this protein that gives red cells their colour.

As blood flows through the alveoli in the lungs, haemoglobin combines with oxygen to form **oxyhaemoglobin**. When blood flows through respiring tissues, oxyhaemoglobin splits up into haemoglobin and oxygen. The oxygen that is released is used by cells for respiration.

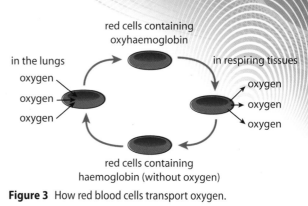

Figure 3 How red blood cells transport oxygen.

Artificial blood

Scientists have been developing different types of artificial blood since the 1990s. This is to overcome the shortage of blood available for transfusions. Artificial blood could also overcome some of the problems associated with transfusing natural blood. Artificial blood could be sterilised to prevent infections being transmitted. It would avoid problems with mismatching of blood groups and it could be stored for long periods.

Most types of artificial blood are known as **HBOCs** (haemoglobin-based oxygen carriers). Some HBOCs are already in use, e.g., in South Africa where there is a chronic shortage of natural blood for transfusions because of the AIDS epidemic. HBOCs are made from haemoglobin, but doctors cannot transfuse haemoglobin itself into patients because it quickly disintegrates once transfused. There are several ways of preventing this disintegration. The most common is to bind the haemoglobin to a synthetic polymer. HBOCs do not usually remain in the blood for more than a day, compared with the 100-day life of a red blood cell.

A second type of artificial blood contains **PFCs** (polyfluorocarbons). PFCs are entirely synthetic. They are efficient oxygen carriers; they can carry 20–30 per cent more oxygen than plasma. However, they can only be used if the patient is supplied with extra oxygen. PFCs are much smaller than red blood cells. They can be used to supply oxygen to places that cannot be accessed by red blood cells, for example through swollen brain tissue. HBOCs and PFCs still produce some side effects.

Route to A*

Practice evaluation questions like the one below. You will be assessed on using good English, organising information clearly and using specialist terms where appropriate.

Use the information given here on artificial blood to evaluate the use of artificial blood for blood transfusions.

Questions

1 Explain why blood looks red.

2 Which type of cell is most common in the blood?

3 **(a)** Give two differences between a red blood cell and a white blood cell. **(b)** Give two differences between a white blood cell and a platelet.

4 List the substances transported by plasma.

5 Explain what is meant when we say 'haemoglobin combines reversibly with oxygen'.

6 Only one red blood cell at a time can pass along a capillary. Explain how this increases the efficiency of diffusion of oxygen into body cells.

7 EPO is a drug that increases the number of red blood cells in the bloodstream. It is used to treat patients with severe anaemia. Athletes are banned from using this drug because it increases their level of performance. Why does taking EPO enhance athletic performance?

8 Explain how the structure of red blood cells is adapted for their function.

Stents, artificial heart valves, artificial hearts

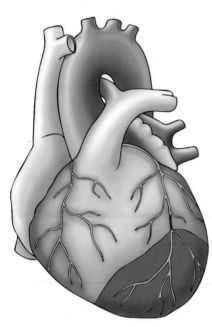

Figure 1 The coronary blood vessels run over the surface of the heart, carrying blood to and from the heart muscles.

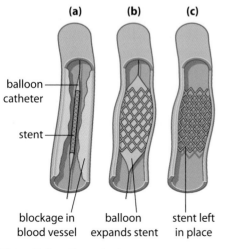

Figure 3 Inserting a stent.

Surgery and stents

Heart muscle needs oxygen to contract efficiently. Oxygenated blood is carried to the heart muscle by the coronary arteries, which branch off the aorta.

In coronary heart disease, layers of fatty material build up inside the coronary arteries and cause them to narrow. This reduces the flow of blood through the coronary arteries. As a result, the heart muscles get less oxygen. In severe cases, the flow of blood to part of the heart muscle ceases and the affected muscle tissue dies; this is what is known as a heart attack.

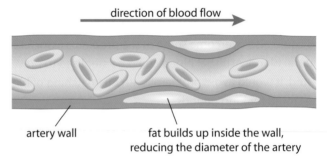

Figure 2 Fat deposits in the wall of a coronary artery.

There are two methods of treating coronary artery disease: bypass surgery and the use of stents.

In bypass surgery the surgeon uses a piece of a vein from the leg, chest or arm to create a bypass around the blocked portion of coronary artery. The operation requires a general anaesthetic and up to seven days in hospital. It takes up to three months to fully recover.

A stent is a wire metal mesh tube used to prop open an artery. The stent is collapsed to a small diameter and put over a balloon attached to a narrow tube called a 'catheter'. The catheter is inserted into a blood vessel in the leg and manipulated into the area of the blockage in the coronary artery. Then the balloon is inflated, the stent expands, locks in place and forms a scaffold. This holds the artery open. The catheter is then removed. The stent stays in the artery permanently, holds it open and improves blood flow to the heart muscle. However, fatty substances may build up in stents over the years.

Inserting a stent takes 1–2 hours and is carried out under a local anaesthetic. Most patients stay overnight in the hospital for observation, and can resume normal activities within one week. However, fatty substances are not as likely to build up in a bypass as they are in stents.

Replacing heart valves

In most people heart valves operate faultlessly throughout life, but in a few the heart valves can become faulty. There are two main faults:

- the heart valve tissue might stiffen, preventing the valve opening fully
- the heart valve might develop a leak.

Faulty heart valves can be replaced using biological valves taken from humans or other mammals, or by mechanical valves. Figure 4 shows the two types.

Modern mechanical valves consist of two semicircular carbon leaflets that pivot on hinges. The leaflets swing open completely, parallel to the direction of the blood flow.

Mechanical valves are very strong and can last a lifetime, which is why they are often used in young patients. However, mechanical valves damage red blood cells and increase the risk of blood clotting. To minimise this risk, patients have to take drugs that prevent blood clotting for the rest of their lives.

The most common operation using biological valves is to replace the valve in the aorta with the patient's own pulmonary valve, then to replace the pulmonary valve with a valve from a human donor or from a pig's heart. Biological valves do not damage red blood cells, but they have a tendency to become hardened with calcium deposits. This means that after a while they may not open fully.

Artificial hearts

Some patients with heart disease cannot be treated by using stents or artificial heart valves. For many of these patients there are only two options: a heart transplant or an artificial heart. Artificial hearts are still at the experimental stage and have not been very successful so far.

Figure 5 shows one type of total heart replacement. This artificial heart requires both external and internal power supplies.

biological valve
(from human or other mammal)

mechanical valve

Figure 4 Artificial heart valves.

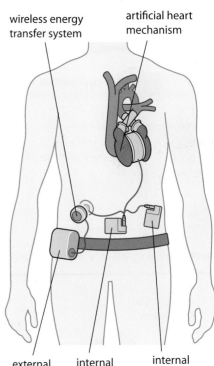

Figure 5 An artificial heart.

Science skills

A study was made of 81 patients at high risk of death due to irreversible heart failure. The patients were given an artificial heart as an interim treatment while waiting for a heart transplant. The rate of survival to transplant operation was 79 per cent, compared with 46 per cent in a group of control patients who did not receive an artificial heart. The one-year survival rate among patients who received the artificial heart was 70 per cent, compared with 31 per cent among the controls. The one- and five-year survival rates among patients who received transplants were 86 per cent and 64 per cent, respectively.

a Use information from the diagram and the text to evaluate the use of artificial hearts to treat heart failure.

Questions

1 Evaluate bypass surgery and stent use as methods of treating blocked coronary arteries.

2 Evaluate the use of mechanical and biological valves as methods of treating heart failure.

Transport systems in plants

Learning objectives

- explain the role of xylem in transporting water around a plant
- explain the role of phloem in transporting sugars around a plant.

Movement of water through a plant

The concentration of **solutes** in water around a plant's root system is lower than the concentration of solutes in the cytoplasm of root hair cells, so water moves into these cells by osmosis (see lesson B3 1.1). This makes their cytoplasm more dilute than the cytoplasm of the cells further inside the root, so water moves from the root hair cells into these cells by osmosis. In this way water travels across the root until it reaches xylem tissue in the centre.

In the leaves, water evaporates from the cells that line the air spaces. The water vapour then diffuses out of the leaf through the stomata and into the air. As the water evaporates, this increases the concentration of the cytoplasm in the cells lining the air spaces, so water moves by osmosis into these cells from the cells that are next to them. This happens from cell to cell across the leaf until the xylem in the leaf vein is reached. There, water moves by osmosis from the vein into the cells that surround it.

Most of the water travels through xylem vessels. These are formed by the breakdown of the end walls of dead xylem cells, forming long tubes. Xylem vessels function like the water pipes in your home. They have thick, rigid walls that stop them from collapsing.

As water moves out of xylem cells in a leaf, the water columns in the xylem vessels are pulled upwards because water molecules stick together. Water is literally pulled all the way from the roots to the top of the plant. This is called the transpiration stream. An enormous amount of energy is needed to do this in tall trees such as redwoods. This energy comes directly from the Sun, which causes water to evaporate from the leaves.

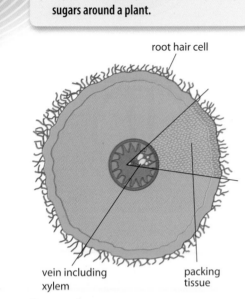

root hair cell

vein including xylem

packing tissue

Figure 1 A cross-section across a root.

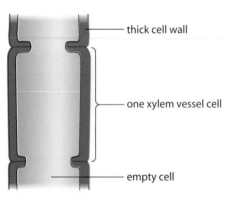

thick cell wall

one xylem vessel cell

empty cell

Figure 2 Xylem vessels carry water throughout a plant.

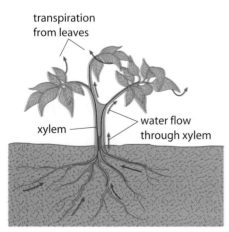

transpiration from leaves

xylem

water flow through xylem

Figure 3 Movement of water through a plant.

Vascular bundles

Another plant tissue called 'phloem' transports sugars in plants. Phloem tissue is always found next to xylem tissue in **vascular bundles**. Figure 4 shows the position of xylem and phloem in different parts of a plant.

Most of the sugar transported by phloem is carried through sieve tubes. These are similar to xylem vessels, but the end walls of the phloem cells do not break down completely. Instead, they form structures called 'sieve plates'. Mature sieve tubes are not empty, but they have no nucleus. They are always associated with living cells called 'companion cells'. The activity of these companion cells is essential for the activity of phloem.

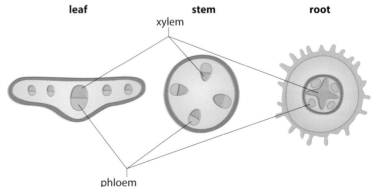

leaf　　　　stem　　　　root

xylem

phloem

Figure 4 Xylem and phloem in leaves, stems and roots.

Sugars are produced in leaves by photosynthesis. This sugar is then transported to all parts of the plant. All plant cells need sugar for respiration. Plants produce new cells at the tips of roots and stems. These new cells need sugars to grow as well as to respire. Both root tips and shoot tips are supplied with sugars by phloem.

Most plants store large amounts of carbohydrates in underground stems or roots. For example, potatoes store starch and sugar beet stores sugar.

An Italian scientist called Marcello Malphighi first showed that phloem transported sugars, over 300 years ago. He removed a ring of bark from a tree. The bark of a tree contains the phloem. After a few weeks a swelling appeared above the ring. The swollen tissue was rich in sugars. This suggested to Malpighi that sugars were being transported down the tree from the leaves to the roots. He repeated the experiment in winter. This time no swelling appeared. Malpighi correctly concluded that this was because in winter there were no leaves to produce sugars, so no sugars were transported down to the roots.

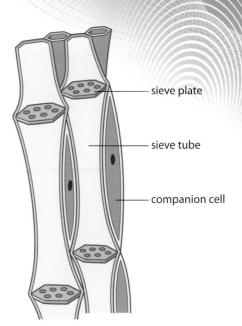

Figure 5 Sieve tubes.

Scientists have found that the movement of sugars through phloem is much faster than can be explained by diffusion. They have also found that if they poison companion cells, movement of sugar in the phloem stops. Their conclusion is that companion cells assist in the movement of sugars through active transport.

Giant redwoods can move water up to a height of 100 m.

Questions

1 Name the plant tissue that transports the following. **(a)** Mainly water. **(b)** Sugars.

2 In which direction do these occur? **(a)** Water moving through a plant. **(b)** Sugar moving through a plant.

3 **(a)** Describe the structure of a xylem vessel. **(b)** Describe the structure of a sieve tube.

4 Explain how water passes from xylem across a leaf.

5 Explain what causes water to move to the top of a tall tree.

6 Describe the location of vascular bundles in the following parts of plants. **(a)** Roots. **(b)** Stems. **(c)** Leaves.

7 Explain how xylem vessels are adapted to carry out their function.

8 Compare and contrast the mechanisms by which water and sugars are moved through plants.

Examiner feedback

Remember the different energy sources that are used for transporting substances through the plant.

Movement through the xylem depends on thermal energy from the Sun evaporating water from the leaves. Movement through phloem depends on energy from respiration. So transport through xylem does not require living cells, whereas movement through phloem requires energy released by respiration in mitochondria.

Science in action

Scientists now use radioactive tracers to follow the movement of sugars around a plant. They label carbon dioxide with a **radioactive isotope** of carbon. The plant takes up this radioactive carbon dioxide and uses it to make sugars. These radioactive sugars can be tracked around the plant using radiation detectors such as Geiger counters or photographic film.

Assess yourself questions

1 **(a)** Explain what is meant by diffusion. *(2 marks)*

(b) Figure 1 shows four ways in which molecules may move into and out of a cell. The dots show the concentration of molecules

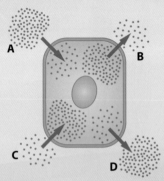

Figure 1 Movement of molecules in and out of a cell.

Which arrow, A, B, C or D, represents the movement of:

(i) carbon dioxide during photosynthesis? *(1 mark)*

(ii) carbon dioxide during respiration? *(1 mark)*

2 A student set up an experiment. She used discs of potato, each cut to the same size. In batches of five, she dried the discs on paper towel then weighed them. She put one batch of discs into each of five beakers with the following contents:

- beaker 1 – distilled water
- beaker 2 – 10% sucrose solution
- beaker 3 – 20% sucrose solution
- beaker 4 – 30% sucrose solution
- beaker 5 – 40% sucrose solution.

After two hours she dried the discs then reweighed them.

(a) **(i)** What was the independent variable in this experiment? *(1 mark)*

(ii) What was the dependent variable? *(1 mark)*

(iii) Give two ways in which the student tried to improve the reliability of her experiment. *(2 marks)*

(b) Table 1 shows the student's results.

(i) Copy and complete the table. Two sets of calculations have been done for you. *(3 marks)*

Table 1 Student's results.

	Beaker 1	Beaker 2	Beaker 3	Beaker 4	Beaker 5
Initial mass/g	9.9	10.5	10.0	10.1	10.4
Final mass/g	13.0	12.2	9.1	8.0	7.4
Change in mass/g	+3.1				−3.0
Change in mass (%)	+31.3				−28.8

(ii) Explain why the student calculated percentage change in mass rather than just change in mass. *(2 marks)*

(iii) What type of graph should the student draw to display her results? Explain the reason for your choice. *(2 marks)*

(iv) Draw a graph of the results. *(4 marks)*

(c) Give one way in which the student could have improved the precision of her results. *(1 mark)*

(d) Explain why the discs gained mass in beaker 1. *(2 marks)*

(e) **(i)** Use your graph to find the concentration of sugar solution in which potato discs would not change in mass. *(1 mark)*

(ii) Explain why potato discs would not change mass in this solution. *(2 marks)*

3 Figure 2 shows some alveoli and a blood capillary in the lung.

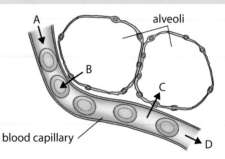

Figure 2 Alveoli and a capillary.

(a) Which of the letters A–D shows:

(i) oxygenated blood?

(ii) the diffusion of oxygen?

(iii) the diffusion of carbon dioxide? *(3 marks)*

(b) Describe two features of the alveoli that help gas exchange. *(2 marks)*

(c) Explain how oxygen that diffuses into the blood is transported around the body. *(3 marks)*

4 Figure 3 shows the structure of a root hair cell.

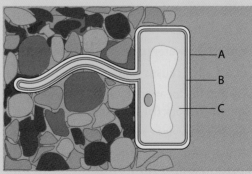

Figure 3 A root hair cell.

(a) Which of the structures A–C is partially permeable?
(1 mark)

(b) Explain what is meant by the term 'partially permeable'.
(1 mark)

(c) Use the diagram to describe two features of the root hair cell that make water uptake more efficient.
(2 marks)

(d) Explain how water is absorbed from the soil into the root hair.
(3 marks)

5 Four leaves were removed from the same plant. Petroleum jelly (a waterproofing agent) was spread on some of the leaves, as follows:

• leaf A – on both surfaces

• leaf B – on the lower surface only

• leaf C – on the upper surface only

• leaf D – none applied.

The leaves were placed near a window and weighed at intervals. The results are shown in Figure 4.

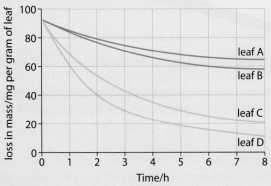

Figure 4 Mass loss from four leaves.

Each surface of the leaf was observed using a microscope. Figure 5 shows the appearance of the upper and lower surfaces of the leaves.

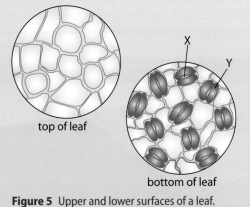

Figure 5 Upper and lower surfaces of a leaf.

(a) What process causes the loss in mass? *(1 mark)*

(b) From which surface is the most mass lost? Use the results of the investigation to explain the evidence supporting your answer. *(3 marks)*

(c) Suggest why the units to show the loss in mass were in mg per gram of leaf. *(1 mark)*

(d) Name space X. *(1 mark)*

(e) Name cell Y. *(1 mark)*

(f) Use the diagram to explain the difference in the results obtained for leaves B and C. *(2 marks)*

6 Figure 6 shows a simplified plan of the human circulatory system.

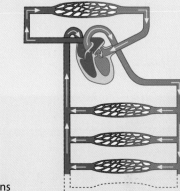

vein

artery

capillaries in lungs

capillaries in body organs

Figure 6 The human circulatory system.

(a) Sketch the diagram and draw appropriate lines from the labels to the diagram. *(2 marks)*

(b) Explain how you chose where to label the artery. *(1 mark)*

(c) In which kind of blood vessel, artery, capillary or vein, does most exchange of substances occur? *(1 mark)*

(d) Explain how this kind of blood vessel is adapted for exchange. *(1 mark)*

7 Figure 7 shows a section through the heart.

Figure 7 A cross-section of the heart.

(a) Copy the diagram and label the following.

(i) The left ventricle. *(1 mark)*

(ii) The vena cava. *(1 mark)*

(iii) The aorta. *(1 mark)*

(iv) A valve that prevents blood returning to the heart. *(1 mark)*

(b) Give one advantage of having a double circulatory system. *(1 mark)*

Staying in balance

B3 3.1

Learning objectives

- describe some internal conditions in the body that are homeostatically controlled
- describe processes in which waste products are excreted from the body
- explain why temperature, blood sugar levels, water and ion content are controlled in the body.

In and out

We are constantly taking substances into our bodies, through our lungs and in what we drink and eat. We also make many substances in our bodies during chemical reactions. Some of these substances are waste products that could damage cells or interfere with other reactions if too much collects in the body.

These waste products must be **excreted**, that is, removed from the body. For example, carbon dioxide produced during **respiration** is excreted, mostly through the lungs when we breathe out. Another waste product is **urea**, made from excess **amino acids**, which is excreted through kidneys (see lesson B3 3.2).

We are also continually losing other substances from our bodies when we sweat and breathe, and in **faeces** and **urine**. This continual exchange with the environment means that conditions inside our bodies could change quite rapidly. However, it is essential that this doesn't happen.

The need for balance

There are many processes in the body that keep the key substances and the body temperature within quite narrow limits. These processes are known as **homeostasis**.

Food and drink don't just contain the sugars, proteins and fats that get broken down and absorbed into your body. They also provide water and **mineral ions**, such as the sodium ions and chloride ions that make up salt. These are all small molecules and easily pass across cell membranes. Most cells, including muscle and nerve cells, will only function properly if there is the right balance of ions, such as sodium and potassium ions, on either side of the cell membrane.

Too much water in your cells can cause them to swell and even burst. Too little water can cause cells to shrink and function less efficiently. Lack of water also makes it difficult for other substances to move around. The movement of water in and out of cells is linked to the ion content in the cells. If the ions are not at the right concentration, then the water may move into or out of the cells by **osmosis** (see lesson B3 1.1). Simple sugar molecules, such as **glucose**, can also affect the movement of water by osmosis. Glucose is carried around the body in the blood, so blood glucose concentration must also be kept within limits.

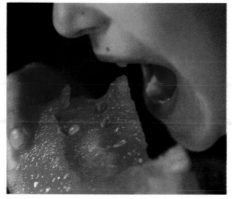

There are many substances in our food and drink, including water, many ions and complex chemicals.

Marathon runners must drink during a race, or they will become **dehydrated**. Drinking a lot of pure water can cause the body to absorb too much water into cells, causing overhydration, which is also dangerous.

Science in action

People with cholera or dysentery suffer severe diarrhoea and lose water and ions rapidly. Rehydration salts, dissolved in clean water, provide the right balance of ions and water to replace these losses. This cheap but very effective medicine has saved millions of lives.

You have seen earlier that **enzymes** only work properly within quite a limited range of temperature and pH. Enzymes control so many of the chemical reactions in our cells that we quickly become unwell if these conditions change too much. Carbon dioxide dissolves easily and makes the solution more acidic, which would affect cell reactions if the carbon dioxide was not removed.

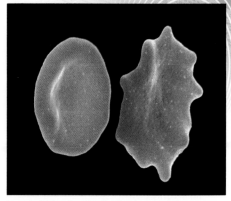

Two red blood cells: the one on the left is normal, the one on the right has been in a solution that contained a lot of mineral ions.

Science skills

a Describe the effect of temperature on the rate of reaction.

b In view of the information on enzyme activity in the graph, explain why it is important that body temperature in humans is kept constant at about 37 °C.

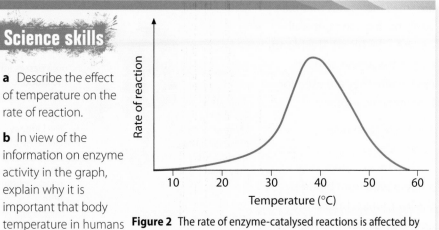

Figure 2 The rate of enzyme-catalysed reactions is affected by temperature.

Route to A*

When discussing homeostasis, and the effect of changes on the body, remember the effects of temperature and pH on the active site of enzymes. Be prepared to discuss any changes in the body in terms of the importance of enzymes in the control of many body processes, and therefore the key role of homeostasis in keeping the body functioning normally.

Questions

1 Write a definition of the term *homeostasis*.

2 Describe and explain the difference between the two red blood cells in the photographs.

3 Suggest what would happen if the red blood cell on the left was put into pure water. Explain your answer.

4 For each of the following, say if the person is likely to become dehydrated or overhydrated. Give a reason for each answer. **(a)** Jenny doesn't have a drink all day. **(b)** Denny drinks two litres of water in one go. **(c)** Benny eats a large pack of salted crisps. **(d)** Penny sits outside for two hours on a hot summer day.

5 Cholera is a disease that causes lots of watery diarrhoea and vomiting and can cause death in a matter of hours. Explain why the death rate from cholera is high and why treatment with rehydration salts dissolved in clean water greatly reduces the death rate.

6 Breathing rate in the body is controlled by homeostatic mechanisms that are more sensitive to carbon dioxide concentration in the blood than to oxygen concentration. (Note that when carbon dioxide dissolves in blood it forms an acidic solution.) Explain as fully as you can why it is important that control of breathing rate involves carbon dioxide receptors. It may help to think about what would happen if carbon dioxide builds up in the cells.

Taking it further

Ions play a far more important role in cells than just maintaining osmotic balance. The transfer of sodium ions across cell membranes facilitates the movement of glucose into cells. Cells are also continually exchanging sodium for potassium, to keep cell volume under control. This is particularly important in nerve cells, where the ion exchange across membranes produces nerve impulses. The sodium–potassium pumps in cell membranes account for about one-third of all the cell's energy use.

The role of the kidneys

Learning objectives

- describe how urea is produced from the breakdown of excess amino acids
- explain how healthy kidneys produce urine by reabsorbing glucose, dissolved ions and water needed by the body, and releasing urea and excess water and ions
- outline the role of the kidneys in homeostasis
- explain the value of urine tests for diagnosis in forensics and medicine.

Examiner feedback

Remember that it is the liver that makes urea and the kidneys that excrete urea.

Examiner feedback

You will not be expected to know the structure of tubules for your exam, but you will need to understand the role of the processes of filtration and reabsorption in producing urine.

The production of urea

During the digestion of your food, proteins are broken down to amino acids. These are small molecules that pass easily through the gut wall and into the blood. The blood carries the amino acids to cells where they are used for building new proteins.

The body cannot store excess amino acids, so these have to be excreted. They are removed from blood by the liver, which breaks the amino acids down to make a substance called urea. Urea is also a small molecule, so it diffuses easily into the blood. It must be excreted from the body because it is **toxic** in high concentrations and could harm cells and tissues. It is cleared from the blood in the **kidneys**.

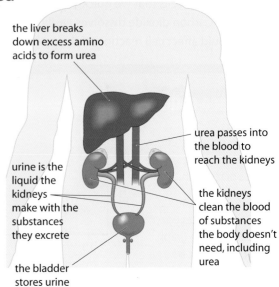

the liver breaks down excess amino acids to form urea

urea passes into the blood to reach the kidneys

urine is the liquid the kidneys make with the substances they excrete

the kidneys clean the blood of substances the body doesn't need, including urea

the bladder stores urine

Figure 1 How excess amino acids are changed to urea, which is then excreted from the body.

Making urine

You have two kidneys about halfway down your back and inside your abdomen. Each kidney is made up of over one million tubules. The role of the **tubules** is to filter your blood and remove some of the substances your body doesn't need, including urea.

At the start of each tubule is a small capillary network. The cells that line these capillaries have very leaky membranes. This allows a lot of **plasma** (the liquid part of the blood), and the small molecules dissolved in it, to be filtered from the blood, leaving the cells and large molecules behind in the blood vessel.

Each tubule is closely associated with capillaries. As the fluid passes along the tubule, many of the substances in it are absorbed back into the capillary; all the sugar (glucose) is reabsorbed, for example. This is important because the body uses glucose for respiration. Also, any dissolved ions and water that the body needs are reabsorbed. This makes sure that the right balance of water and ions is maintained in the body so that cells and all the processes in them can work properly.

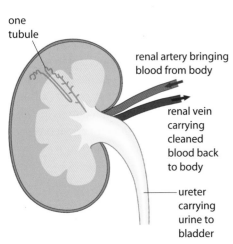

one tubule

renal artery bringing blood from body

renal vein carrying cleaned blood back to body

ureter carrying urine to bladder

Figure 2 Section through a kidney showing its blood supply.

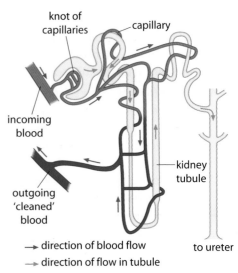

knot of capillaries

capillary

incoming blood

outgoing 'cleaned' blood

kidney tubule

→ direction of blood flow
→ direction of flow in tubule

to ureter

Figure 3 One tubule and its capillary.

The fluid that remains in the tubule is called **urine**. It passes to the bladder where it is stored until it is excreted.

The water that is reabsorbed moves back into the blood by osmosis. Some substances are reabsorbed passively from the tubule into the blood by diffusion. Others are actively reabsorbed by active transport, which uses energy.

Science skills

Table 1 shows the concentrations of different substances in the blood, in the liquid filtered off into the kidneys, and in urine.

Table 1 Concentrations of different substances in the blood, kidney filtrate and urine.

Substance	Concentration / grams/100 cm³		
	blood plasma	kidney filtrate	urine
glucose	0.10	0.10	0.0
mineral ions	0.43	0.43	0.18
protein	6.5	0.0	0.0
urea	0.02	0.02	2.0
water	91.0	91.0	96.0

a Explain the similarities and differences in concentration between the blood plasma and the filtrate.

b Which substance is reabsorbed in the tubule?

c Suggest what would happen to the concentrations in urine of these substances on a hot day.

d Suggest why protein is not part of the filtrate.

Questions

1 Suggest some of the substances dissolved in plasma that could pass into the tubule.

2 Which structures in blood cannot pass into the tubule? Explain your answer.

3 Draw a flow chart to show how excess amino acids are removed from your body.

4 If you drink the same amount of liquid, you will produce less urine on a hot day than on a cold day. Explain why.

5 A doctor may test a patient's urine for glucose. Explain what glucose in the urine might indicate.

6 The reabsorption of glucose and ions in the kidneys uses a combination of diffusion and active transport. Explain fully why both processes are involved.

7 Ethanol (alcohol) is a small molecule that diffuses easily into the blood from the gut and is not actively reabsorbed in the kidneys. Explain why a urine test is an accurate way of measuring blood alcohol concentration and why urine samples must be taken from drivers at the scene of a car accident.

8 Explain fully the role of the kidney in homeostasis.

Science in action

Urine tests are a quick and simple way of identifying what is going on chemically in the body. Doctors use them for diagnosing diabetes and a wide range of other conditions. The police use urine tests to estimate how much alcohol drivers have been drinking. Urine tests can also be used to find out if someone is pregnant.

Practical

When alcohol is added to potassium dichromate, the colour changes from yellow to green. This simple test can identify alcohol concentration in urine.

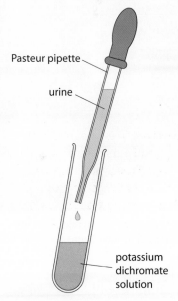

Pasteur pipette

urine

potassium dichromate solution

Figure 4 A urine test for alcohol.

Treatment with dialysis

Learning objectives

- describe how a person with kidney failure may be treated with dialysis and explain why it has to be carried out at regular intervals
- explain how substances are exchanged between dialysis fluid and the patient's blood across partially permeable membranes to restore normal levels in the blood
- evaluate the advantages and disadvantages of treating kidney failure by dialysis.

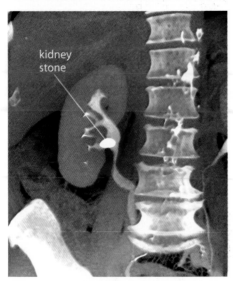

Kidney stones form when salts in the urine crystallise out to make large lumps that block the kidney tubes.

Examiner feedback

Remember that substances in solution diffuse from regions of high concentration to regions of lower concentration.

Route to A* A*

There is no active transport in a dialyser, so exchange of substances across the membrane is passive only. This is important to remember when contrasting dialysis with the way kidneys work, as it explains the concentrations of substances in the dialysis solution.

Kidney failure

The kidneys play an important role in homeostasis of the body. If they fail to work properly, it can soon cause problems. The concentration of urea in the body increases, and water and ions get out of balance. Kidneys can fail to work properly if they have been damaged in an accident or by disease.

Many things can cause kidney disease, such as infection, diabetes, long-term high blood pressure, or simply blockage of the tubes by hard lumps called kidney stones. Many people have kidney disease, sometimes without even knowing it, because a little damage has little effect on kidney function.

Only a small proportion of people with kidney disease will eventually develop **kidney failure**. This is defined as occurring when the kidneys function at less than 30 per cent of their normal level. At this stage the body can be affected badly by the lack of kidney function. The treatment offered by a doctor will depend on the cause of the disease. For example, kidney stones can be broken up by ultrasound, but other kinds of damage cannot be mended so easily.

Without treatment, a patient with kidney failure will eventually fall into a coma and die from the toxins in the blood, or from a heart attack.

Dialysis treatment

One way of treating kidney failure is **haemodialysis**. This is when a machine takes over the function of the kidneys. Needles are inserted into blood vessels, often in the patient's arm. Blood flows from the patient through a tube to the machine and is returned through another tube. Inside the machine the blood flows through a filter called a **dialyser**.

Inside the dialyser, a partially permeable membrane separates the blood from the dialysing fluid. This fluid contains glucose and useful minerals. It has similar glucose and ion concentrations to those in normal plasma.

If the blood is low in glucose or minerals, these will diffuse into the blood from the dialysing fluid. If their concentration is too high in the blood, then the excess will diffuse out into the dialysing fluid. Waste products, such as urea, diffuse out of the blood, because there are none of these substances in the dialysing fluid.

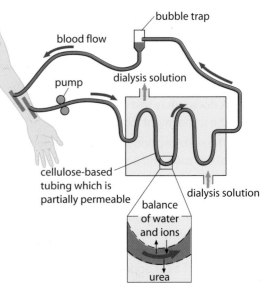

Figure 1 Diffusion into and out of the blood happens in a dialyser just like in the kidney.

Practical

We can model the way dialysis works using visking tubing to help understand the process.

a How would you set up the experiment using the apparatus in Figure 2? What would you put in the boiling tube? What would you put in the visking tubing?

b The visking tubing model has limitations when used to explain the mechanism of dialysis. What are these limitations?

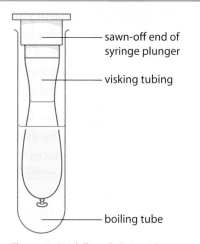

- sawn-off end of syringe plunger
- visking tubing
- boiling tube

Figure 2 Modelling dialysis with visking tubing.

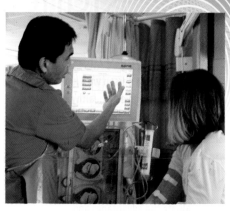

A nurse checks that a patient's haemodialysis is going well.

How good is dialysis?

Dialysis only partly copies the action of the kidneys, so people with kidney failure also need to control their diet to prevent a build-up of particular substances, and to avoid taking in too much fluid. Haemodialysis is usually carried out three times a week, for about four hours each time, and often in hospital with medical supervision. Peritoneal dialysis is a similar treatment that is performed continually at home so the patient can move around at the same time. However, patients and their families have the responsibility of making sure the dialysis is carried out properly.

Examiner feedback

You do not have to remember the terms haemodialysis or peritoneal dialysis.

Questions

1 Suggest why people can have kidney disease without knowing it.

2 Explain why a stone blocking urine from leaving the kidney could affect how a kidney works.

3 How is dialysis useful in treating kidney failure?

4 What are the drawbacks of using dialysis for treating kidney failure?

5 Explain how the dialysis machine mimics the normal function of the kidney, and the ways in which it does not mimic the normal function of the kidney.

6 Explain in detail what will happen to the concentration of substances in the blood such as urea and digestion products if the kidneys are failing. What effect will this have on the cells in the body?

7 Explain as fully as possible why people who are on dialysis must control what they eat.

8 Compare the advantages and disadvantages of haemodialysis and peritoneal dialysis, and suggest different groups of people for whom each method would be more suitable.

Peritoneal dialysis uses a natural membrane in the abdomen as the dialysis membrane.

Controlling blood glucose

Eating foods that contain lots of sugar or starch soon raises the blood glucose concentration.

Taking it further

When glucose is taken into liver and muscle cells it is converted to glycogen, which is a large carbohydrate, rather like starch. It is insoluble and does not affect the osmotic balance of the cell, so it is a good storage molecule. When blood glucose concentration falls, and glucagon is released, glycogen is quickly converted to glucose and released into the blood.

Glucose in the blood

When you **digest** starchy or sugary foods, they are broken down to simple sugars, mostly glucose, and absorbed into your blood. If all the glucose stayed in your blood it would be dangerous as it causes water to move out of cells by osmosis. The body has to take glucose out of the blood so it can be used for respiration or stored until it is needed.

Normal blood glucose concentration varies throughout the day, but it is kept within limits by another feedback control mechanism in the body. This mechanism is monitored and controlled by the **pancreas**, a digestive organ situated behind the liver.

Controlling blood glucose

If the blood glucose concentration gets too high, the pancreas releases a **hormone** called **insulin**, see Figure 1. Insulin circulates in the blood and causes cells, particularly in the liver and muscles, to take glucose out of the blood. Insulin release into the blood decreases as blood glucose concentration comes back down to normal.

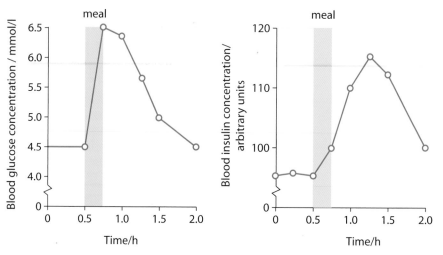

Figure 1 Changes in concentration of blood glucose (left) and insulin (right) after a meal.

If blood glucose concentration drops too low, for example during vigorous exercise, another hormone, called **glucagon**, is released from the pancreas. Glucagon stimulates liver cells to release stored glucose into the blood so that blood glucose concentration rises again.

Diabetes and blood glucose concentration

Some people cannot control their blood glucose concentration properly. We say they have a disease called **diabetes**. Symptoms of diabetes include excreting glucose in urine, unusual thirst and lack of energy. If blood glucose concentration stays high, it can lead to coma and even death, so it is essential that it is brought back down again.

A simple and quick test for diabetes is to check for glucose in the urine using Clinistix. The diagnosis is then checked more thoroughly using a glucose tolerance test. In this test, the patient is given a glucose drink, and then samples of blood are tested for blood glucose concentration to see how quickly the glucose is controlled.

A patient was given 50 g of glucose to drink at time 0. Blood glucose concentration was measured at half-hourly intervals and compared with results from a healthy person and a person who was severely diabetic.

Table 1 Glucose tolerance test results.

Time/hours	Blood glucose concentration / mg/1000 cm³		
	healthy person	severely diabetic	patient
0.0	92	247	108
0.5	124	262	159
1.0	145	323	212
1.5	118	304	150
2.0	95	278	129

a Explain the changes in the patient's results over time.

b From the results, do you think the patient is diabetic? Explain your answer.

Types of diabetes

There are different causes of diabetes and they are treated in different ways. In **type 1 diabetes**, the pancreas produces no insulin at all, so blood glucose concentration can rise to dangerous levels after eating. Treatment requires injections of insulin, balanced with a healthy, controlled diet and exercise, to keep blood glucose concentration within a safe range.

Examiner feedback

For the exam you need to know about the causes, symptoms and treatment of type 1 diabetes, but you do not need to know about type 2 diabetes.

Questions

1 Compare the role of the hormones insulin and glucagon in keeping blood glucose concentration within limits in a healthy person.

2 Before a patient is given a glucose tolerance test, they must not eat or drink anything other than water for several hours. Why is this?

3 Explain why the presence of glucose in the urine can be used to diagnose diabetes.

4 Draw a feedback diagram like the one in lesson B3 3.5 to show how blood glucose concentration is kept within limits in a healthy person.

5 Explain why a Clinistix test is followed with a glucose tolerance test to confirm diabetes.

6 Explain in detail why exercise affects blood glucose concentration.

7 Explain fully what diabetes is and why it must be treated.

8 Explain as fully as you can the shape of the graphs of blood glucose concentration and insulin concentration in Figure 1.

Practical

To test for glucose in the urine, dip a stick into the urine to wet it completely and wait for the colour on the stick to develop.
The colour is compared with a chart to indicate how much glucose is present.

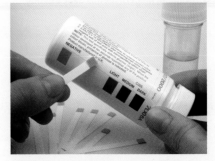

Clinistix to test for glucose in urine.

Taking it further

The other main type of diabetes is type 2. In type 2 diabetes, either the pancreas doesn't make enough insulin or the body cells don't respond properly to insulin. For most people, a healthy, controlled diet and aerobic exercise are sufficient treatment. However, some need tablets to help the body make more insulin or respond better to insulin.

The proportion of people who have type 2 diabetes is increasing, and many doctors think this is related to the increase in the proportion of people who are **obese**.

Treating diabetes

Learning objectives

- describe different methods of treating diabetes
- explain the effect of different methods of treating diabetes on patients
- evaluate modern methods for treating diabetes.

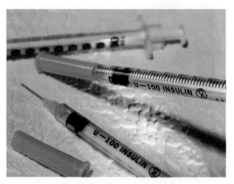

Insulin is injected just below the skin, for example in the stomach or thigh, and diffuses into blood vessels.

The effects of diabetes

The problems with diabetes are not just extreme swings in blood glucose concentration. Over many years, poor control of blood glucose can cause damage to small blood vessels all over the body. In the eyes, this damage can lead to a form of blindness. Diabetes can also cause kidney damage. Up to one-third of patients may need a kidney transplant after suffering years of diabetes.

The better blood glucose concentration is controlled, the less damage is caused.

Controlling blood glucose

In addition to exercise and a healthy diet, patients with type 1 diabetes need treatment with insulin. Insulin is usually injected because it is a protein and would therefore be broken down in digestion if taken by mouth.

Different forms of insulin are used: some are fast acting and injected just before every meal; others are slow acting and injected once a day. Many patients use a combination of slow-acting insulin to produce a background level of insulin all day and fast-acting insulin to control the rapid change in blood glucose concentration after a meal. This produces control that is more like the body's own process.

Some diabetics now use **insulin pumps**, which continually supply insulin just under the skin. Most of the time this is at a background level, but the dose is easily increased at mealtimes.

Science skills
The results in Table 1 come from a survey of 272 diabetic patients who used either an insulin pump or multiple daily injections to control blood glucose concentration.

Table 1 Comparison of insulin pump with multiple insulin injections.

Measure	Insulin pump (continual delivery)	Multiple insulin injections
mean blood glucose concentration / mmol/l	7.45	7.67
fluctuation in blood glucose concentration / mmol/l	±3.9	±4.3

a Normal blood glucose concentration varies between about 4 and 8 mmol/l. Which method gave the best control of blood glucose? Explain your answer.

Each method of treating diabetes has problems, and some problems will affect some patients more than others. In the UK, patients can only be given a pump if they pass certain criteria that suggest they will get greater benefit using a pump than using injections.

US golfer Kelly Kuehne is a diabetic who wears an insulin pump on her waist. This photo shows her in the 1999 Women's US Open Golf Championship.

Table 2 Advantages and disadvantages of different methods of treating diabetes with insulin.

Method	Advantages	Disadvantages
multiple insulin injections	• discreet: injection syringe or pen can be carried around in a bag and used in privacy • equipment is cheaper	• greater chance of extreme high or low blood glucose concentration • uses more insulin each day than pump
insulin pump	• better control of blood glucose concentration • uses less insulin per day	• must be worn almost all the time • equipment more expensive

Measuring the insulin dose

Before insulin is injected, the correct **dose** must be calculated. The dose will depend on the concentration of blood glucose, and how it is expected to change in the next hour or so, such as at a mealtime. So diabetics have to take a blood test every time they need an injection. This can now be done quickly and simply using a pinprick of blood measured with a blood glucose meter.

Using a blood glucose meter is easy, even with children.

Science in action

Scientists are developing automatic blood glucose meters combined with an insulin pump that can be implanted below the skin, so that the process is completely automatic and discreet.

b Explain the advantages of a completely automatic monitoring and delivery system.

c Suggest a disadvantage of a completely automated system.

Questions

1 Explain why insulin cannot be taken in tablet form.
2 In the US, patients have to pay for much of their medication. What impact do you think this may have on the number of diabetics using an insulin pump? Explain your answer.
3 Explain the importance of having a simple system for testing blood for glucose.
4 Explain why doctors use criteria to help decide who gets which treatment.
5 Explain how a mixed treatment of injections of fast-acting and slow-acting insulin controls blood glucose more like the natural way than treatment using just one form of insulin.
6 In a study of insulin pumps, one teenage girl who spent a lot of time on the beach with her friends chose to go back to injections. Suggest why she preferred injections.
7 Explain fully why blood glucose concentration must be measured before an insulin dose is given. Include the effect of exercise and food in your answer.

Route to A*

When comparing different treatments for type I diabetes, remember to explain their advantages and disadvantages fully in terms of keeping blood glucose concentrations as closely within the normal range as possible. It is the extremes of blood glucose concentration that cause damage to cells, and the nearer to normal blood glucose concentrations can be maintained, the more healthy the patient is likely to remain.

ISA practice: rates of transpiration

A number of factors affect the rate of transpiration. If the rate of transpiration is too high, plants may wilt.

A horticulturist has asked students to investigate the effect of wind on the rate of transpiration. The horticulturist's hypothesis is that the air movement will affect the rate of transpiration.

Section 1

1   Write a hypothesis about how you think wind affects the rate of transpiration. Use information from your knowledge of physical processes, such as evaporation, to explain why you made this hypothesis.   *(3 marks)*

2   Describe how you could carry out an investigation into this factor.

   You should include:

   - the equipment that you could use

   - how you would use the equipment

   - the measurements that you would make

   - how you would make it a fair test.

   You may include a labelled diagram to help you to explain the method.

   In this question you will be assessed on using good English, organising information clearly and using specialist terms where appropriate.   *(6 marks)*

3   Think about the possible hazards in the investigation.

   (a) Describe one hazard that you think may be present in the investigation.   *(1 mark)*

   (b) Identify the risk associated with the hazard that you have described, and say what control measures you could use to reduce the risk.   *(2 marks)*

4   Design a table that you could use to record all the data you would obtain during the planned investigation.   *(2 marks)*

   Total for Section 1: 14 marks

Section 2

A groups of students, Study Group 1, carried out an investigation into the hypothesis. They used an electric fan to simulate wind. They measured the mass of a plant shoot in different 'wind' speeds for thirty minutes. Their results are shown in Figure 1.

Zero

Wind speed 1 m/s: mass 75.6 g

Wind speed 2 m/s: mass 76.3 g

Wind speed 3 m/s: mass 75.1 g

Wind speed 4 m/s: mass 75.4 g

After 10 minutes

Wind speed 1 m/s: mass 75.5 g

Wind speed 2 m/s: mass 76.0 g

Wind speed 3 m/s: mass 74.8 g

Wind speed 4 m/s: mass 74.9 g

After 20 minutes

Wind speed 1 m/s: mass 75.3 g

Wind speed 2 m/s: mass 75.7 g

Wind speed 3 m/s: mass 74.4 g

Wind speed 4 m/s: mass 74.3 g

After 30 minutes

Wind speed 1 m/s: mass 75.1 g

Wind speed 2 m/s: mass 75.5 g

Wind speed 3 m/s: mass 74.0 g

Wind speed 4 m/s: mass 73.8 g

Figure 1 Study Group 1's results.

5   **(a)** Plot a graph of these results.   *(4 marks)*

   (b) What conclusion can you make from the investigation about a link between air movement and the rate of transpiration? You should use any pattern that you can see in the results to support your conclusion.   *(3 marks)*

   (c) Look at your hypothesis, the answer to question 1. Do the results support your hypothesis? Explain your answer. You should quote some figures from the data in your explanation.   *(3 marks)*

   Here are the results of three more studies.

   Below are the results from three other study groups.

   Figure 2 shows the results from Study Group 2, a second group of students.

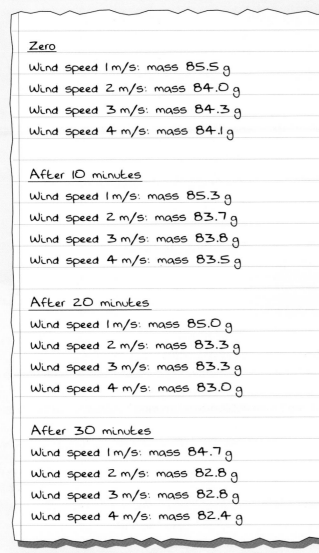

Zero

Wind speed 1 m/s: mass 85.5 g

Wind speed 2 m/s: mass 84.0 g

Wind speed 3 m/s: mass 84.3 g

Wind speed 4 m/s: mass 84.1 g

After 10 minutes

Wind speed 1 m/s: mass 85.3 g

Wind speed 2 m/s: mass 83.7 g

Wind speed 3 m/s: mass 83.8 g

Wind speed 4 m/s: mass 83.5 g

After 20 minutes

Wind speed 1 m/s: mass 85.0 g

Wind speed 2 m/s: mass 83.3 g

Wind speed 3 m/s: mass 83.3 g

Wind speed 4 m/s: mass 83.0 g

After 30 minutes

Wind speed 1 m/s: mass 84.7 g

Wind speed 2 m/s: mass 82.8 g

Wind speed 3 m/s: mass 82.8 g

Wind speed 4 m/s: mass 82.4 g

Figure 2 Study Group 2's results.

Table 1 shows the results from Study Group 3. This group of students studied the effect of a different factor on the rate of transpiration.

Table 1 Results from Study Group 3.

Humidity (%)	Transpiration rate/ arbitrary units
20	26.0
40	21.0
50	16.5
60	11.0
70	9.5

Study Group 4 was a group of researchers, who looked on the internet and found a graph (Figure 3) showing the effect of a third factor on the rate of transpiration.

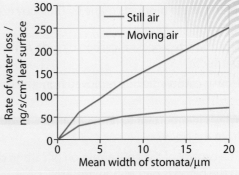

Figure 3 Graph showing the effects of another factor on transpiration.

6 **(a)** Draw a sketch graph of the results from Study Group 2. *(3 marks)*

(b) Look at the results from Study Groups 2 and 3. Does the data support the conclusion you drew in answer to question 5(c)? Give the reasons for your answer. *(3 marks)*

(c) The data contain only a limited amount of information. What other information or data would you need in order to be more certain whether or not the hypothesis is correct?
Explain the reason for your answer. *(3 marks)*

(d) Look at the results from Study Group 4. Compare it with the data from Study Group 1. Explain how far Study Group 4's results support or do not support your answer to question 5(b). You should use examples from Study Group 4 and Study Group 1. *(3 marks)*

7 **(a)** Compare the results of Study Group 1 with Study Group 2. Do you think that the results for Study Group 1 are *reproducible*?
Explain the reason for your answer. *(3 marks)*

(b) Explain how Study Group 1 could use results from other groups in the class to obtain a more *accurate* answer. *(3 marks)*

8 Suggest how ideas from the original investigation and the other studies could be used by the horticulturist to reduce the risk of his plants wilting. *(3 marks)*

Total for Section 2: 31 marks

Total for the ISA: 45 marks

Assess yourself questions

1 One of the waste products made by the body is carbon dioxide.

  (a) From which process is carbon dioxide a waste product? *(1 mark)*

  (b) How is carbon dioxide removed from the body? *(1 mark)*

  (c) Why is it important to remove carbon dioxide from the body? *(2 marks)*

  (d) The control of the concentration of carbon dioxide in the blood is part of homeostasis. Define homeostasis. *(2 marks)*

  (e) Name three other conditions in the body that are homeostatically controlled. *(3 marks)*

2 Some students were wondering why marathon runners are wrapped in foil blankets at the end of a race. One thought it was to help lose heat faster because metal is a good conductor of heat. Another student thought it was to do with survival. They carried out an experiment using the equipment in Figure 1.

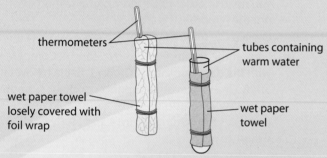

thermometers

tubes containing warm water

wet paper towel losely covered with foil wrap

wet paper towel

Figure 1 Apparatus to test effectiveness of foil in reducing heat loss.

  (a) Suggest why approximately 40 °C is a suitable starting temperature for the water. *(1 mark)*

  (b) Explain why both tubes were wrapped in wet paper towels. *(1 mark).*

  (c) Describe one systematic error that this experimental set-up makes possible. *(1 mark)*

  (d) Explain how this error could be avoided. *(1 mark)*

  (e) Table 1 shows the results from the experiment (measurements were made to the nearest 0.5 °C). Graph the data using a suitable chart format. *(3 marks)*

  (f) Using your graph, identify the anomalous results and suggest one way in which they could have happened. *(2 marks)*

  (g) Write a conclusion for this experiment using these results. *(3 marks)*

  (h) Using what you know about the control of body temperature, explain as fully as possible why marathon runners are often wrapped in foil blankets after a race. *(4 marks)*

3 Table 2 shows the proportion of certain substances in blood plasma and urine under normal conditions.

Table 2 Percentage in blood and urine.

	Blood (%)	Urine (%)
Protein	8	0
Urea	0.03	2
Glucose	0.1	0

  (a) Where is urine formed? *(1 mark)*

  (b) Which two key processes produce urine from blood plasma? *(2 marks)*

  (c) Explain the difference in proportion of protein and urea in blood and in urine. *(2 marks)*

  (d) A doctor uses Clinistix to measure glucose in this person's urine. State the condition the doctor is testing for, and explain fully why this test is a good indicator for the condition. *(3 marks)*

4 Figure 2 shows a patient receiving dialysis.

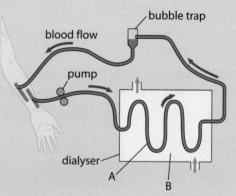

bubble trap

blood flow

pump

dialyser

A

B

Figure 2 Diagram of a patient on dialysis.

  (a) Explain why this patient needs dialysis. *(2 marks)*

Table 1 Heat loss from two test tubes.

Time/min	0	2	4	6	8	10	12	14	16	18	20
Temperature/°C: wet-wrapped tube	39.5	37.0	35.0	34.0	30.0	28.0	27.0	25.5	24.5	23.5	22.5
Temperature/°C: foil-wrapped tube	39.5	38.5	36.5	36.0	33.0	31.0	30.0	29.0	28.0	27.0	26.0

(b) What does line A represent, and what properties does this have in the dialyser?
Explain your answer. *(3 marks)*

(c) The yellow area labelled B is the dialysing fluid. Copy and complete Table 3 to show what is in this fluid before dialysis begins. Explain your choices. *(4 marks)*

Table 3 Substances in dialysing fluid.

Substance	Present	Not present
glucose		
sodium ions		
urea		
water		

(d) Explain why the dialysing fluid is continually flowing through the dialyser. *(1 mark)*

(e) Describe *one* advantage for this patient of receiving dialysis. *(1 mark)*

(f) Describe *two* disadvantages for this patient of receiving dialysis for their condition. *(2 marks)*

5 Figure 3 shows the change in blood glucose (sugar) concentration for a person over one day.

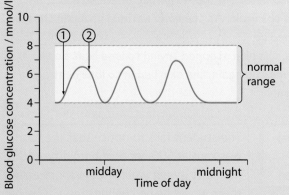

Figure 3

(a) Suggest the cause of the increase in blood glucose concentration at point 1 on the graph. *(1 mark)*

(b) Explain as fully as you can why the blood glucose concentration is falling at point 2. *(3 marks)*

(c) Explain why blood glucose concentration is normally kept within a limited range in the body. *(1 mark)*

(d) On another day, this person went to the gym at midday for an hour of vigorous exercise.
   (i) Sketch a graph to show how their blood glucose concentration would have changed over that hour. *(1 mark)*
   (ii) Explain fully why you would expect their blood glucose concentration to change like that. *(3 marks)*

6 Figure 4 shows the blood flow through the skin in certain conditions.

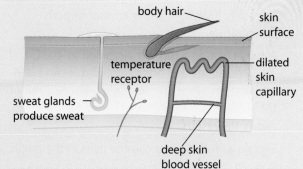

Figure 4 Cross-section through the skin.

(a) In which conditions does the skin respond like this? Explain your answer. *(2 marks)*

(b) Explain the role of skin temperature receptors in the control of body temperature. *(1 mark)*

(c) Describe three changes in the skin that you would expect to see if core body temperature fell below 37 °C and explain fully what effect these changes have. *(3 marks)*

(d) A student attaches a temperature sensor to her left hand and puts her right hand into a bowl of very cold water.
   (i) What effect, if any, will the sensor on the left hand record? *(1 mark)*
   (ii) Explain as fully as you can why this happens. *(3 marks)*

7 An alternative method for treating kidney disease is to give the patient a kidney transplant.

(a) Describe the *main* advantage of a transplant compared with receiving dialysis. *(1 mark)*

(b) Give *one* reason why not all kidney disease patients are suitable for receiving a transplant. *(1 mark)*

(c) Explain as fully as you can the problems with kidney transplants and how those problems are minimised. Your answer should refer to the immune system. *(5 marks)*

8 Modern methods of treating type 1 diabetes include self-injections of insulin by the patient and the automatic insulin pump.

Compare the *advantages* and *disadvantages* of these two methods of treatment.

Use your knowledge and understanding of the natural control of blood glucose concentration in your answer.

In this question you will be assessed on using good English, organising information clearly and using specialist terms where appropriate. *(6 marks)*

Here are three students' answers to the following long-answer 6-mark question.

Scientists investigated the control of body temperature.

A subject sat in a room maintained at 40 °C and 100% humidity. His brain temperature and skin temperature were both recorded. After 25 minutes the subject drank some iced water. Figure 1 shows the results.

Explain as fully as you can the data in the graph.

In this question you will be assessed on using good English, organising information clearly and using specialist terms where appropriate.   (6 marks)

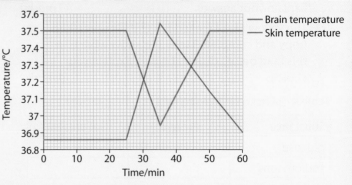

Figure 1 Changes in brain and skin temperature after drinking iced water.

Read the answers together with the examiner comments. Then check what you have learnt and try putting it into practice in any further questions you answer.

B Grade answer

Student 1

Cold water is absorbed into the blood in the stomach/small intestine, lowering the temperature of the blood.

There is no reference to respiration as the source of the heat.

The temperature of the brain dropped after the iced drink because iced water entered the blood, which cooled the blood to the brain. As the body warmed up again the blood warmed up and thus the brain warmed up.
When the subject swallowed the ice, the skin stopped trying to lose heat so the skin temperature rose. As the body warmed up again, the skin began to lose heat again and the skin temperature went down.

Avoid phrases such as 'the skin was trying to lose heat'. This implies that the skin has a will of its own.

There was no reference to the normal body temperature, nor mechanism involved: in this case vasoconstriction.

The student fails to notice that the room was kept at 100 per cent humidity. At this humidity, sweat would not evaporate and there would be no cooling.

Examiner comment

There is an attempt to explain the data for both skin and brain temperatures but there is a lack of clarity and detail. The answer has some structure and organisation. The use of specialist terms is limited.

MOVING UP THE GRADES

Read the question carefully.
- Before beginning, mark on the graph the main changes in direction. Number them, and make sure that you write a short paragraph about each numbered change.
- Note that you are asked for an *explanation*. No marks are awarded for a description, so do not waste time describing each graph.

- Next, decide which mechanisms are involved in the temperature changes. Always read the introduction to the data. In this case, the information that the room is maintained at 40 °C and 100% humidity is crucial.
- Jot down a list of relevant technical terms to include in your answer, e.g., hypothalamus, vasoconstriction, vasodilation.

A Grade answer

Student 2

A good introduction and an adequate explanation for the cooling of the brain. Correct use of terminology – 'circulation' and 'absorption'.

Correct use of terminology: 'capillaries' and 'radiated'. However, there is no reference to constriction of the blood vessels supplying these capillaries.

The brain temperature decreased as cooler blood circulated through it. The blood was cooled as the iced water was absorbed into the blood from the stomach and the small intestine.

The skin temperature rose because the amount of blood flowing through the surface capillaries was reduced. This meant that less heat radiated from the skin surface, resulting in less cooling of the skin surface. Also, the room was at 40 °C, so some heat would transfer from the room to the skin.

Since the blood flowing through the skin was no longer being cooled as much, the blood temperature increased, increasing the brain temperature. As the brain temperature rose, blood flow through the surface capillaries increased and the skin began to lose heat.

A good explanation of the decrease in skin temperature, but it would have been improved by reference to vasodilation and the role of the hypothalamus.

Examiner comment

There is a clear, balanced and fairly detailed explanation of the changes in temperature, but details such as vasoconstriction/vasodilation and the role of the hypothalamus are lacking.

The answer is coherent and in an organised, logical sequence. It contains some specialist terms that are used accurately.

A* Grade answer

Student 3

Good use of terminology in 'monitored and controlled', 'hypothalamus', 'impulses', and 'temperature receptors'.

Good reference to role of hypothalamus.

An excellent account of vasoconstriction. Make sure that you understand this term. Never refer to blood vessels moving up and down in the skin, or the capillaries getting wider or narrower.

Body temperature is monitored and controlled by the hypothalamus in the brain. The hypothalamus has receptors sensitive to the blood flowing through it. The hypothalamus also receives impulses from temperature receptors in the surface of the skin.

Absorption of cold water from the stomach reduces the temperature of the blood. This cooler blood reduces the temperature of the brain as it circulates. The hypothalamus sends impulses to the blood vessels supplying the skin capillaries, causing them to constrict. This reduces the rate of blood flow through the surface capillaries, and so less heat is lost via radiation. The reduced heat loss and the higher temperature of the room cause the skin temperature to increase.

The increase in skin temperature is detected by receptors, which send impulses to the hypothalamus.

The hypothalamus sends impulses to the blood vessels supplying the skin surface capillaries, causing them to dilate. The increase in the rate of blood flow through the surface capillaries increases the amount of heat loss via radiation and the skin temperature begins to fall.

Examiner comment

There is a detailed explanation for each of the changes in temperature. The explanations are coherent and in an organised, logical sequence, based on the monitoring and controlling

functions of the hypothalamus. The answer contains a large range of relevant specialist terms, all of which are used accurately.

Humans and the environment

The human population continues to increase at the expense of other species. In some natural ecosystems, such as oceans or tropical rainforests, because of over-fishing or deforestation the balance of populations may shift. Some species find it difficult to survive. To supply the estimated human population of 9.5 billion people by 2050 with enough food and clean water, changes in agriculture will be needed. More cereal crops instead of leafy vegetables and soft fruit will need to be grown. Cattle rearing may also need to be reduced, limiting our choice of food. We may have to rely on local fish near the bottom of the food chain, rather than those caught in oceans far away. Biotechnology industries can supply protein-rich food cheaply from fungi. Positive and negative effects of managing food production and distribution must be evaluated.

Human settlements around the world generate waste as solids, liquids or gases. Each category of waste can cause permanent damage to the local and global environment. Scientific data related to environmental issues, such as pollution and reduction of biodiversity, must be analysed and interpreted. Some managed waste can be fermented to produce biogas for use as a fuel.

Ethanol and biodiesel are also biofuels. Using them may cut carbon emissions from transport vehicles or power stations. However, to grow the plant stock, like sugar cane, wheat or palm oil, to produce the biofuels requires land, which may have been deforested to grow crops. If wheat or sugar cane is fermented for transport fuel there is less for human food and prices rise. Compromises must be reached between competing priorities.

Test yourself

1 What are food webs made up of and what do arrows in a food web represent?

2 Explain how the knowledge of food webs can help predict the effect of altering the number of various organisms in it.

3 List three renewable energy sources and explain what they can be used for.

4 Name two gases that pollute the air and the main source of each gas.

Objectives

By the end of this unit you should be able to:

- evaluate the methods used to collect environmental data, and forecast future climate trends
- interpret scientific data about the environmental effects of producing biofuels
- describe a variety of biogas generator designs, suited to their location
- explain how mycoprotein is produced in an industrial fermenter
- analyse data about food production to assess its efficiency
- evaluate the positive and negative effects of intensive animal farming
- explain the importance of maintaining breeding levels in fish stocks
- describe methods needed to secure global food and freshwater supply for an increasing population.

Human activities produce waste

Learning objectives

- describe how the human population is growing and explain why increasing amounts of waste are being produced from human activity
- analyse how increased waste may pollute different areas of the environment.

The downside of increased living standards and human population growth

The March 2010 world population estimate is over 6.8 billion. Six thousand years ago the population of the world was about 0.2 billion. People lived in small groups and most of the world was unaffected by human activities.

Many people in the UK have money to spend on non-essential goods and services. Shopping to replace items like televisions and mobile telephones with the latest technology, or furniture and clothes with the latest styles, generates waste and uses up the world's resources. Manufacturing, transport and packaging all contribute to more waste. Our standard of living is increasing but this increases demands on raw materials to build better roads, houses, schools and hospitals.

Science skills

Look at Figure 1.

a Describe how the world's population changed between 1750 and 2000.

b What percentage of the estimated world population in 2050 will live in less developed countries?

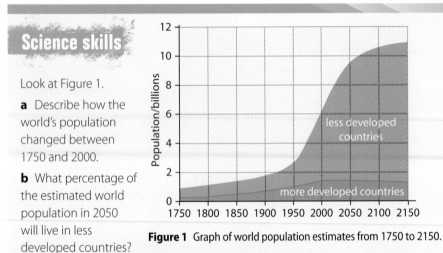

Figure 1 Graph of world population estimates from 1750 to 2150.

c Suggest reasons why the populations of less developed countries are rising faster than those of more developed countries.

Household waste: recycle, reuse, incinerate or bury

Some waste can be recycled and used to make new products so we don't use more raw materials. However, it costs money and uses energy to collect and sort all the waste before it can be recycled.

In 2008–09, household waste accounted for 89% of local authorities' waste collection. However, the proportion of household waste recycled, reused or composted continued to increase between 1996 and 2008 in England (see Figure 2). The government has given local authorities recycling targets to meet.

Most household waste is dumped in landfill sites. As buried waste decays, it produces methane gas. This can be piped away and used to heat homes and offices, or burned to produce heat for generating electricity. Toxic liquids can also be released from landfill sites and may pollute water. The amount of rubbish dumped in landfill tips has decreased in recent years.

The new generation of refuse incinerators removes toxic chemicals and small particles from the smoke and fumes produced by burning. Most modern incinerators are 'energy from waste' plants, which use the heat from incineration to generate electricity. The ash from incinerators must be disposed of safely, or it can cause land or water pollution.

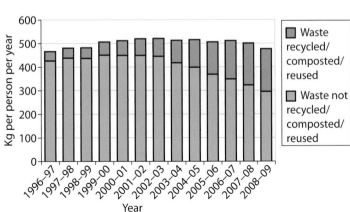

Figure 2 Household waste and recycling in England 1996–97 to 2008–09.

Preventing pollution

Unless the waste we produce is properly handled, it may pollute water. Sewage from our homes and industries, fertiliser used in gardens and on farms, and toxic chemicals from landfill and industry can end up polluting water. Whole ecosystems are affected and take many years to recover.

The air may be polluted by smoke or by gases such as sulfur dioxide (SO_2). Sulfur dioxide is a waste product from some power stations, from industrial processes and from vehicles that use fossil fuels. Some scientists have predicted that global SO_2 emissions will reach a peak around 2020, as developing countries, driven by rapid economic development, increase their use of fossil fuel. Sulfur dioxide contributes to acid rain, which can damage forests and freshwater organisms.

Modern farming practices often involve the use of large amounts of chemicals, including **pesticides** and **herbicides**. Some of these chemicals end up in the soil. Some break down quickly into less harmful chemicals, but others may last and harm animals and plants in the soil for many years. These chemicals may also be washed from land into water, and so cause water pollution.

Less space for wildlife

The land used for building cities and other human communities previously provided habitats for plants and animals. Building materials such as brick and cement are made from raw materials such as clay and limestone, which are obtained by quarrying. Quarrying can threaten the habitat of rare species. Farming and dumping waste pollutes land that would otherwise be available for wildlife.

Polluted waste from copper mining. Many organisms are killed when water becomes polluted.

- Chemicals
- Metallic
- Non-metallic
- Discarded equipment
- Animal and plant
- Mixed waste
- Common sludges
- Mineral wastes

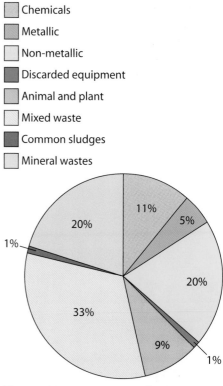

Figure 3 Human waste comes from many sources.

Questions

1 Give two reasons why the production of waste from human activities is increasing annually.

2 How can industrial waste pollute the environment?

3 Describe the risk to the environment from chemicals such as herbicides and pesticides.

4 List the human activities that reduce the amount of land available for other living organisms.

5 **(a)** How much of local authorities' waste collection was accounted for by household waste in 2008–09? **(b)** What government action has been taken to reduce this amount?

6 Look at Figure 2. **(a)** During which years was most household waste produced? **(b)** What is the difference between the not recycled, combusted or reused waste in 1996–97 and 2008–09? **(c)** What percentage of household waste in 2008–09 was recycled? **(d)** Describe the trend in the data for amount of recycled, composted and reused waste from 1996 to 2009.

7 Look at Figure 3. **(a)** Tabulate the data for the following types of waste: chemicals, metallic, discarded equipment, animal and plant, and mineral wastes. **(b)** Display the data in a bar chart.

8 Explain the advantages and disadvantages of disposing of refuse by recycling and incinerating refuse.

Tropical deforestation and the destruction of peat bogs

Tropical deforestation

Despite increased awareness of the importance of tropical rainforests, deforestation rates have not slowed down. Timber is extracted and land is made available for agriculture. Figures from the Food and Agriculture Organization of the United Nations (FAO) show that tropical deforestation rates increased by 8.5 per cent between 2000 and 2005.

Trees are plants: they make their food by photosynthesis. They remove carbon dioxide from the atmosphere and combine it with water to make sugars. The sugars are then used to synthesise the plant tissues. Trees are sometimes described as **carbon stores** because carbon (from carbon dioxide) is locked up in wood and other tissues for many years, until it is broken down either during respiration or when the tree dies. Deforestation reduces the number of trees and other forest plants available to take up carbon dioxide. Microorganisms break down parts of trees that are cut down, and release the carbon dioxide into the atmosphere through respiration. Clearing the deforested area to prepare it for another use often includes burning the tree debris and other vegetation. This increases the release of carbon dioxide into the atmosphere.

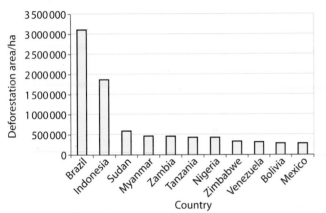

Figure 1 Brazil and Indonesia had the world's highest deforestation rates between 2000 and 2005.

Reduction in biodiversity

Tropical rainforests have abundant rainfall all year round and the temperature is warm. High humidity, warmth and sunlight provide ideal growing conditions, so the trees grow very tall. There is also incredible diversity. In any one hectare of tropical forest you may find trees of one hundred different species. Each tree supports many other organisms. Plants, such as orchids and vines, and fungi grow on the trees. Animal species are found either in the tree canopy 40 metres above the ground, or on the relatively clear forest floor.

Over 50 per cent of the species on Earth live in rainforests. These species will be lost, and **biodiversity** will be reduced, if rainforests are cleared. Some of these species might be useful in the future for making medicines. Forty per cent of western **pharmaceuticals** come from natural plant or animal products. Habitats and wildlife are destroyed when rainforest is destroyed, but the forests are home to indigenous people too. They lose their land when rainforest is cut down.

Biofuels instead of forests

Crops such as palm oil from Indonesia and Malaysia are grown on land that was previously rainforest. Some oil-powered power stations in the UK use palm oil as a fuel rather than oil from fossil fuel. This is because by 2020, 20 per cent of all energy used in the European Union has to come from renewable sources, and palm oil is a good replacement fuel. Palm oil sells as a fuel at a high price. This has led to some palm oil producers illegally cutting down rainforests to make space for growing more palm oil. Palm oil can be chemically reacted with an alcohol to form biodiesel fuel.

Adding to the atmosphere's carbon dioxide concentration.

Food not trees

Population increases and increased trading in developing countries has led to demand for new farmland to grow more food. In Asia, rainforest has made way for more rice fields. In Brazil, some land has been deforested to provide space and grass for cattle-rearing. In addition to the loss of rainforest, these changes of land use increase methane production. Cattle produce and release methane during digestion. Paddy fields, where rice plants grow in water, produce methane through the rotting of plant material. Methane is a powerful greenhouse gas that has eight times the warming effect of carbon dioxide. Increases in methane production are therefore likely to accelerate climate change.

The rosy periwinkle, a native of Madagascar, provides the main drug for treating childhood leukaemia.

Science skills

Table 1 shows the changes in the amount of methane in the atmosphere between 1989 and 2009.

a Display the data as a graph.

b Describe the trend in the data.

c These data were collected in the southern hemisphere. Suggest how the validity of the data could be improved.

Table 1 Changes in atmospheric methane concentration.

Year	Methane concentration in air/parts per billion
1989	1655
1993	1700
1997	1719
2001	1742
2005	1747
2009	1758

Peat bogs and compost

Peat bogs provide unique wildlife habitats and are valuable stores of carbon. There is more carbon held in peat bogs globally than in the world's forests. Extraction of peat releases carbon dioxide into the atmosphere. Some UK peatlands are now protected, but many in Europe are not.

Are biofuels more important than rainforests?

Questions

1 Describe the ways in which deforestation can increase the amount of carbon dioxide in the atmosphere.

2 'Tropical rainforests are the most diverse habitats on Earth.' Provide three pieces of evidence to support this statement.

3 Explain why a reduction in biodiversity is a cause for concern.

4 Name a biofuel that can be used in place of fossil fuels for electricity generation. How can this fuel be converted to biodiesel?

5 How has agriculture led to an increase in methane in the atmosphere?

6 Name two countries that had high deforestation rates between 2000 and 2005 and suggest two reasons for them allowing this to happen.

7 Suggest three reasons for the action taken by the Unilever Corporation in December 2009 in Indonesia.

8 Explain to a new gardener why he should not use peat in his garden.

PEAT-FREE FOR A GREENER GARDEN
ACT ON CO2

Figure 2 All garden centres in the UK now stock peat-free composts.

Environmental effects of global warming

The greenhouse effect

The Earth receives heat from the Sun. Gases in the atmosphere absorb most of this heat. If they didn't, it would be lost back into space. Carbon dioxide and methane are called **greenhouse gases**. They keep the surface of the Earth warmer than it would be without them. This is known as the **greenhouse effect**.

If the amounts of carbon dioxide and methane in the atmosphere increase, this increases the amount of heat reflected back to the Earth's surface. This makes the Earth even warmer. This increased greenhouse effect is known as **global warming**. Global warming is happening, and there is evidence that at least some of it is due to human activities.

Science skills

a What was the mean air temperature in Antarctica 14 000 years ago?

b Describe how the carbon dioxide concentration in the atmosphere changed from 18 000 years ago to the present.

c Explain how the data from the graph give evidence that carbon dioxide is a greenhouse gas.

d Explain why the graph does not prove that carbon dioxide is responsible for global warming.

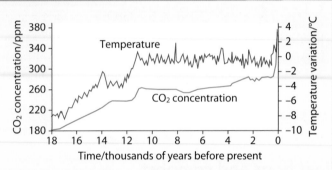

Figure 2 Changes in carbon dioxide levels and global temperature over the last 18 000 years.

Summer visitors to the UK may be reduced if temperatures rise.

Climate change modelling and redistribution of species

Scientists use computer models to predict what will happen to the Earth's climate if global warming continues. There are so many different factors to take into account that there is still a lot of disagreement about what will happen. However, most people think that a rise of only a few degrees could change the way that weather patterns form. This could cause higher rainfall and floods in some places. It would also change where different plants and animals could live. A few scientists feel that more reasearch should be done into possible natural explanations for global warming.

The place where any particular type of tree in the northern hemisphere is found is likely to shift northwards.

Climate change would result in extinctions, and a reduction in biodiversity, as plants and animals fail to either migrate or adapt to changing habitats.

Swallows, house martins and cuckoos visit the UK and northern Europe. They come to nest and reproduce in a cooler climate than Africa, where they spend the rest of the year. Geese and swans visit the UK in winter. If the climate warms, all these birds will need to change their migration times and patterns in order to survive.

Rising sea levels and carbon storage

Increasing temperatures are causing glaciers and polar ice caps to melt and release more water into the oceans. Some scientists predict that this could result in a sea level rise of up to *80 metres*. Millions of people live in areas that would be deep underwater if sea levels rose this much.

Some scientists have proposed using carbon storage or **sequestration** as a way of removing CO_2 from the atmosphere. Several proposals have been made for carbon storage in the oceans, such as depositing CO_2 on the ocean floor in 'lakes' at depths of 3000 metres, where CO_2 is denser than water. However, we do not yet have sufficient scientific understanding to be sure that such ocean storage proposals would work.

Methods used to collect environmental data

To find out about the climate before 1860, scientists have to rely on 'proxy' records rather than direct measurements with instruments. For example, the width of tree rings is related to temperature. Other techniques that have been used include examining the time of crop harvests and other historical records.

Scientists can obtain more reliable data on past temperatures by taking measurements from ice cores. As the ice was formed, air bubbles were trapped in it. Scientists can analyse the air trapped in these bubbles and so measure carbon dioxide levels in the atmosphere hundreds of thousands of years ago. They can also estimate the temperature of the atmosphere when the bubbles were trapped.

The most recent report on climate change produced by the Intergovernmental Panel on Climate Change (IPCC) uses the output from 18 computer models that produce a range of possible scenarios that might result from climate change. Through the media and in the public mind, these scenarios have become predictions.

A scientist examines part of a 3-km ice core drilled from the Antarctic ice cap.

Science skills To improve the reliability of evidence for global warming, measurements are repeated, often many times. If other groups of scientists repeat an experiment and get similar results, this makes any conclusions more valid. Reliability and validity apply both to evidence and to projections of future events.

About 10 per cent of Bangladesh would be flooded if the sea level were to rise by one metre.

Questions

1 Name two greenhouse gases.

2 Explain how an increase in greenhouse gases in the atmosphere causes global warming.

3 Why is the greenhouse effect important for life on Earth?

4 Describe how the distribution of species may be affected by global warming. Give examples of species that may be involved.

5 What changes in sea level do some scientists think could result from global warming? What would be the effects of such a change?

6 Discuss how the media, scientists and the general public perceive the IPCC's climate change scenarios.

7 Give three methods that have been used to collect environmental data from before 1860, when records began. Evaluate two methods and consider their validity and reliability.

8 Evaluate the methods used today to produce scenarios that might result from climate change, and consider their validity and reliability.

Biofuels containing ethanol

Converting crops to biofuels

Carbohydrate-rich crops such as sugar cane, sugar beet, maize and wheat can be **anaerobically fermented** with microorganisms to form ethanol-based fuels. This is known as bioethanol or biofuel because it is produced from living plant material. Starch in the source crop is converted to glucose (hydrolysed) by the action of carbohydrase enzymes. Yeast is added to ferment the sugars anaerobically to carbon dioxide and ethanol. The ethanol can then be separated by distillation and used as fuel instead of, or in combination with, petrol.

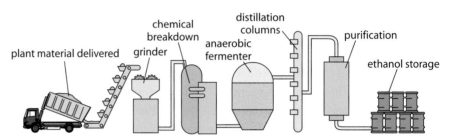

Figure 1 Making bioethanol from plant material. In the fermenter, the plant material is used as food for microorganisms, which produce ethanol as a waste product.

Is ethanol a carbon-neutral fuel?

Ethanol is often described as a 'carbon-friendly' or carbon-neutral fuel. This term is used because the crops grown to make it remove carbon dioxide from the atmosphere as they grow, through photosynthesis. The cars that burn biofuels release carbon dioxide back into the air, but this is only replacing the carbon dioxide that the crops took in. An ideal biofuel produces no net carbon emissions, but in practice bioethanol is not completely carbon neutral. Transport, fertiliser production and processing of the crop to produce ethanol all cause release of carbon dioxide. Biofuels are renewable and do not deplete the Earth's resources. Burning ethanol produces less air pollution than burning petrol.

Using food crops such as wheat, sugar cane and maize to produce bioethanol puts pressure on food prices, causing them to rise, as less land is available to grow food to supply an increasing world population.

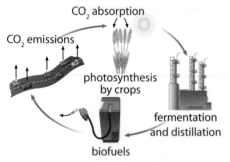

Figure 2 Biofuels emit carbon dioxide, but the plants that they are made from absorb it.

Gasohol is a mixture of ethanol and petrol. In Brazil it contains 22 per cent ethanol.

Table 1 shows the changes in pollutant levels in Brazil for 15 years after the introduction of gasohol. The data show average emissions per car of CO (carbon monoxide), HC (hydrocarbons) and NO (nitrous oxides) in grams per kilometre travelled.

a Why is petrol included in the data?

b Describe the trends in the data.

c Which pollutant shows the greatest percentage decrease since before 1980?

Table 1 Average pollutant emissions after the introduction of gasohol.

Year	Fuel	Pollutant / g/km		
		CO	HC	NO
<1980	petrol	54.0	4.7	1.2
1986	gasohol	22.0	2.0	1.9
	ethanol	16.0	1.6	1.8
1990	gasohol	13.3	1.4	1.4
	ethanol	10.0	1.3	1.2
1995	gasohol	4.7	0.6	0.6
	ethanol	3.2	0.4	0.3

Government directives for renewable energy

By 2020 each European Union (EU) member state must ensure that 10 per cent of total road transport fuels come from renewable energy. Road transport accounts for around 22 per cent of UK carbon dioxide emissions. A world-scale bioethanol facilty in Hull is due to start producing bioethanol in 2010. It will produce 420 million litres of bioethanol each year from UK and EU wheat. Some of the ethanol will be mixed with just 15 per cent petrol to produce a fuel known as E85. 'Flex-fuel' cars have engines that can run on pure ethanol, pure petrol or any mixture of the two. A computer chip analyses the mixture when the tank is filled and adjusts the engine accordingly.

In 2009, flex-fuel vehicles were sold in 18 European countries.

Ethanol obtained from straw and wood

The bar chart in Figure 3 shows the overall carbon dioxide emissions of various fuels used in vehicles. Emissions for each type of fuel can vary, depending how it is produced. However, wood and straw are not used much commercially to produce ethanol. This is because they contain carbohydrate in the form of cellulose, which cannot easily be converted into sugars for fermentation.

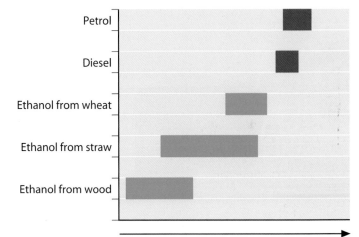

Figure 3 Carbon dioxide emissions from transport fuels.

Questions

1 Explain what is meant by 'biofuels'.

2 **(a)** Describe how ethanol-based fuels are made. **(b)** What are these fuels used for?

3 Explain what is meant by 'carbon-neutral fuels'.

4 Apart from lowering carbon dioxide emissions, give **(a)** two advantages and **(b)** one disadvantage of using biofuels.

5 Explain what gasohol is and how it compares with E85.

6 What is meant by a 'flex-fuel' car and what is the advantage of owning one?

7 Look at Figure 3. **(a)** Which fuel produces the highest carbon dioxide emission? **(b)** Which fuel produces the lowest carbon dioxide emission? **(c)** Which ethanol-based fuel has the highest carbon dioxide emission?

8 Explain how E85 fuels and the gribble worm may help lower the UK's carbon dioxide emissions from transport and why the UK government is encouraging both.

Science in action

In February 2009, researchers at the universities of York and Portsmouth were awarded £2 million as part of the UK's bioenergy research programme. The scientists are working with gribble worms, *Limnoria quadripunctata*, a marine pest that bores through the wood planks of ships and piers. The worm's gut produces enzymes to break down woody cellulose and turn it into energy-rich sugars. Fermentation of the sugars obtained in this way from wood and straw could produce ethanol.

Biogas from small-scale anaerobic fermentation

Microorganisms produce fuel

Biogas is produced by anaerobic fermentation of plant material or animal waste, such as manure, sewage sludge and kitchen waste. Microorganisms in the waste break down the waste materials in the anaerobic conditions of the fermenter, to produce biogas.

Biogas can only be produced in this way if the waste contains carbohydrates. Many different types of microorganisms are involved in the breakdown of the waste materials into biogas. The fuel or biogas produced is made up mainly of methane, with some carbon dioxide. Biogas makes an excellent fuel for cooking stoves and furnaces. The scale of biogas production can vary from small-scale household production to large commercial biogas generators.

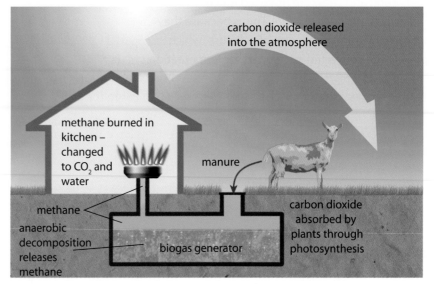

Figure 1 Locally produced biogas can be used for cooking and for heating water.

Biogas for families

The simplest types of biogas generators have no moving parts. Figure 2 shows a biogas generator suitable for a small farm in the developing world. A trench about 10 metres long and 1 metre deep is dug in the ground, and a heavy-duty plastic bag is then placed in the trench. The bag is fitted with inlet and outlet pipes and a valve to draw off the gas. To start up the digester, the bag is two-thirds filled with water, and then topped off with the exhaust gases from a car. Then a mixture of water and faeces is fed through the inlet pipe. It takes about 18 kg of faeces per day to provide enough gas for the farmer's needs. The methane produced is released through the valve at the top. The **effluent** is the solid particles and water left after gas production. The solids settle into a sludge that can be used as a crop fertiliser. The effluent can be run off by lowering the outlet pipe into a ditch.

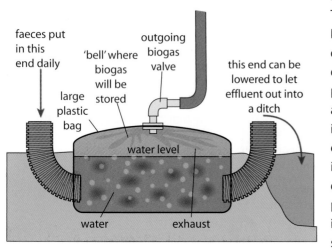

Figure 2 Biogas production in the developing world.

Gas is fed directly to the farmer's house, to be used for cooking and/or heating. The gas can also fuel farm machinery.

Disadvantages of having farm-based biogas generators include having to stock and monitor the biogenerator on a daily basis, using land to house the biogenerator and pipework, and the initial set-up costs. Gas pressure may vary when gas is collected in a simple dome or 'bell'.

In addition to biogas production, biogas generators facilitate nutrient recycling, waste treatment and odour control.

Biogas for rural schools

A biogas generator at a school at Myeka in a remote part of South Africa uses human and animal faeces to produce gas for cooking, science experiments and driving an electric generator. The installation consists of two 20 000-litre digesters. These receive the faeces from 16 school toilets. The toilets are arranged in two concentric circles around the digesters. The installation also incorporates two cow-dung inlet ports. The farm animal waste increases the output from the digesters and allows biogas production to continue at weekends and during holidays when the students are not in school.

The installation at Myeka High School delivers an estimated 13 cubic metres of gas daily. This produces 18 kilowatts of power – enough to run the school's computers and electric lighting. To operate efficiently, biodigesters like the one at Myeka need to be located in a warm climate where there is a good supply of water, so that the fermentation rate of the microorganisms is fast.

What a difference biogas makes.

Constructing a biogas generator similar to the one at Myeka.

Biogas can be used to fuel farm machinery, as on this pig farm in Argentina

Questions

1 What is biogas?

2 What materials are needed to produce biogas?

3 What biological process is used to make biogas?

4 **(a)** Suggest two uses of biogas in a farmer's household. **(b)** Suggest how the sludge remaining after fermentation is used on the farm.

5 How would village hygiene be improved by a biodigester?

6 Where is the gas collected in a simple farm-type of biogas generator? Give an advantage and a disadvantage of this type of collection.

7 Make a table to show the factors that influence the rate at which biogas is produced in a small-scale biogas generator. Alongside each factor explain how it will have its effect.

8 Explain how a school in a remote part of South Africa can produce enough electricity for computers and lighting using biogas.

Generating biogas on a large scale

Using a city's organic waste for biogas

Biogas can be generated on a large scale from manure and organic waste. Figure 1 shows a scheme that operates in Kristianstad in Sweden using waste from several sources (see data box). The biogas is produced in the digester. Most of the gas is piped to a district heating supply in the town.

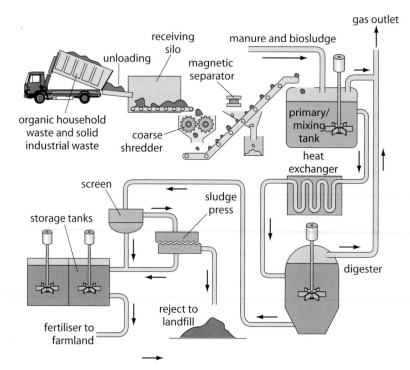

Figure 1 Producing gas from city refuse. The gas is drawn from the primary/mixing tank and the digester.

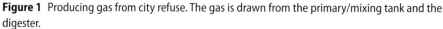

Science skills

Table 1 shows the Swedish biogas digester's annual input, output and energy statistics.

a Calculate the percentage mass of the input material that is converted into biogas.

b Calculate the proportion of the gross biogas production that is needed to heat the plant.

c List the advantages and disadvantages of producing fuel in this way rather than using mains gas.

Table 1 Statistics from biogas digester in Kristianstad in Sweden.

Input/tonnes	
manure	41 200
household waste	3100
abattoir waste	24 600
distillery waste	900
vegetable waste	1400
Output/tonnes	
liquid biofertiliser	67 150
biogas	4000
waste	50
Energy produced/MWh	
gross biogas production	20 000
biogas used to heat plant	2100
biogas sales to district heating plan	17 900
electricity purchased to operate biogas plant	540

Farmers produce biogas on an industrial scale

The town of Tillamook Bay in Canada has constructed a biogas plant to process the manure of 10 000 dairy cows.

The digesters are 40 m × 10 m × 4 m in size. Each digester is equipped with heating and insulation.

Canada is not a warm country. The digester needs both insulation and heating to keep the microbes warm enough to quickly ferment the waste.

This installation cost about £1 000 000 to build. The benefits are:

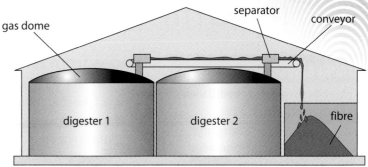

Figure 2 Cow manure digester at Tillamook Bay.

* it provides enough electricity for 150 average homes
* it produces 23 000 tonnes of high-quality potting soil per year
* It returns 64 million litres of liquid fertiliser per year to farmers' fields.

Maximising biogas production

In a biodigester facility some or all of the features shown in Table 2 may be included in the design of the system.

Table 2 The advatages and disadvantages of various digester features.

Feature	Advantage for system	Disadvantage for system
automatic mixing system in feed tank	speeds up digestion in the fermenter	maintenance and energy costs of mixing equipment
pump to force organic matter into the fermenter	reduces labour and supplies a constant amount of feed to the fermenter	maintenance and energy costs of pump
heating system in the fermenter	speeds up the fermentation rate and biogas production in temperate climates	energy costs, especially during cold seasons
agitation system in the fermenter	speeds up the fermentation rate as microbes and organic matter are in closer contact	high energy costs and potential for agitator breakdown
floating gas holder (drum)	constant gas pressure supplied	expensive and requires intensive maintenance

The more efficient the biogas generator is, the more gas and profit it will make. Some features speed up the fermentation rate, while others help regulate the gas supply and pressure.

Questions

1 Give three sources of organic waste for a city biogas reactor other than manure and household waste.

2 What is most of the gas from the Swedish biogas digester used for?

3 Name two secondary products from the gas Canadian production facility supplied by manure.

4 Explain what would happen to the volume of gas output from biogas generators in Sweden and Canada if they were not supplied with heating and insulation.

5 Name two design features of a biodigester facility that help regulate the gas supply and pressure. What are the advantages of these features?

6 If the Tillamook Bay facility was located in the south of France, why would running costs be reduced?

7 Consider the two main types of gas in biogas. Suggest how the waste from a sugar beet factory in Norfolk could optimise growing conditions in the greenhouses of a neighbouring tomato grower.

8 Evaluate the advantages and disadvantages of three design features of biogas generators.

Collecting biogas from landfill and lagoons

Learning objectives

- explain how methane from landfill sites can be collected for use as a fuel
- describe how lagoons containing large volumes of animal waste prevent odour problems and supply biogas
- compare biogas output from a variety of farm animals.

In 2008, 57 million tonnes of waste went to landfill in England and Wales.

Science in action

Some local councils have put radiofrequency identification (RFID) chips into wheelie bins to monitor the amount of landfill waste produced by each household. The idea behind the RFID chips is to increase recycling and decrease landfill.

Biogas from landfill sites

Most waste in the UK is still disposed of in landfill sites. Waste is deposited in a section of the site, and compacted. When the section is full, it is covered with soil. The organic waste dumped in such a landfill site will decompose with time. If there is an impervious surface below the waste, the waste will become waterlogged and anaerobic fermentation will occur. This will lead to production of biogas from the waste, mostly methane. This methane will slowly work its way up through the waste and be vented into the atmosphere. If landfill is covered after use, the gas will slowly seep through the earth covering and be released into the atmosphere.

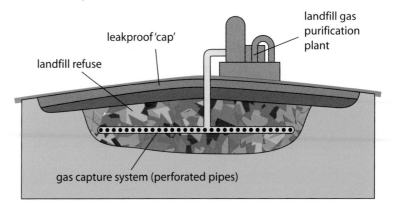

Figure 1 Fuel from rubbish.

Instead of letting methane escape, it can be collected and used as a fuel. Perforated pipes (pipes with holes in) are used to collect the gas. A cover prevents gas escaping into the atmosphere. The production of gas from many landfill sites can continue for around 20–30 years, though there is a gradual reduction after about 10 years.

Biogas from farm lagoons

Farms with large numbers of animals have problems disposing of all the faeces and urine produced. These are often run into large lakes or lagoons and left to decay naturally. The smell can be terrible. However, these lagoons produce biogas, and can be fairly easily adapted to collect it: the lagoon is covered with heavy-duty plastic sheeting to prevent biogas escaping, and tubes are put in to collect the gas. The gas is then used as a fuel on the farm. Biogas lagoons are being built extensively in countries south of the UK, but there are comparatively few in the UK because the climate is too cold.

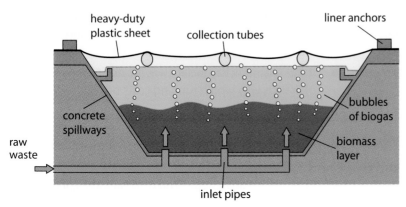

Figure 2 Biogas from an anaerobic lagoon.

Table 1 shows how much biogas and energy we can get from various types of animals and their manure.

Table 1 Biogas and energy from various animals.

	Hen	Pig	Horse	Sheep	Cow	Unit of measurement
mass of each animal	2	70	400	60	500	kg
Manure						
	0.19	5.88	20.40	2.40	43.00	kg per day
	0.085	0.084	0.051	0.040	0.086	kg manure per kg animal per day
Output						
total biogas	0.0167	0.3338	1.62	0.1975	1.24	m^3/day
total power	0.0045	0.0904	0.44	0.0535	0.34	kW
biogas/kg manure	89.43	56.77	79.44	82.28	28.91	litres/kg manure per day
power/kg manure	24.22	15.38	21.51	22.28	7.83	W/kg manure per day
biogas/kg animal	7.60	4.77	4.05	3.29	2.49	litres/kg animal per day
power/kg animal	2.06	1.29	1.10	0.89	0.67	W/kg animal per day

a Display the data for total kg manure per kg animal per day as a bar chart.

b Produce a pie chart to show production of biogas/kg animal as a portion of the total biogas/kg animal from all five animals.

c i What is the general relationship between biogas/kg animal and the mass of the animal? **ii** Are there any anomalies in the relationship you identified in (i)?

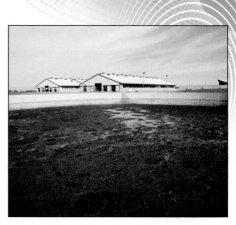

Preventing pollution and generating gas.

Examiner feedback

Remember that many different types of microorganism are involved in the production of biogas. These microorganisms are present in the feedstock material from plants and animals.

Science in action

Landfill gas can be hazardous and can cause explosions. To prevent this, the gas used to be burned on site at the top of a flare stack. Land gas purification plants that capture the gas avoid this waste of energy.

Questions

1 Why is it important to stop methane escaping into the atmosphere?

2 How is methane collected from landfill sites for use as a fuel?

3 Explain what an anaerobic lagoon is.

4 Give two advantages of covered lagoons on farms.

5 Using the biogas statistics above, discuss why an egg production unit would benefit from contributing to a biogas generator.

A*

Protein-rich food from fungus

Science in action

Food for the future?

A 500 kg bullock can yield 1 kg of new protein per day; 500 kg of soya beans can yield 40 kg of new protein per day; but 500 kg of microorganisms could yield 50 000 kg of protein per day.

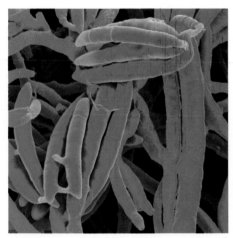

A photomicrograph of the fungus *Fusarium*. Mycoprotein is made from the species *Fusarium venenatum*.

Examiner feedback

Mycoprotein is produced from a micoscopic fungus grown in a fermenter. It is an aerobic process and so cannot be described as a fermentation, which is an anaerobic process.

Fusarium is a soil fungus

The global population is increasing rapidly. This has resulted in increased demand for protein for both people and animals. Using microorganisms as a protein source is one solution to this problem. The fungus *Fusarium venenatum* is used in 50-metre-high fermenters to produce a protein-rich food called mycoprotein. It is suitable for vegetarians because it has no connection whatsoever with animals.

Fusarium lives naturally in the soil, where it feeds on the dead remains of plants and animals. Its fungal body consists of microscopic, narrow, branched, thread-like structures called **hyphae**. These microscopic fungi grow quickly and reproduce, needing relatively little space. In optimum aerobic conditions with a plentiful supply of nutrients, they can double their biomass every five hours.

Growing *Fusarium* in a fermenter

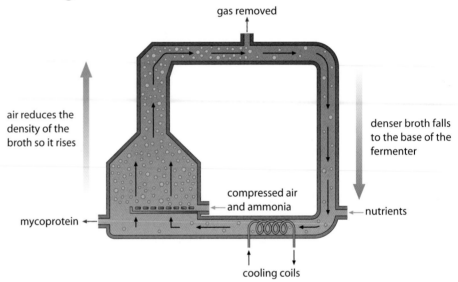

Figure 1 Mixing the contents of the fermenter using compressed air.

Keeping the fungus growing

The contents of the fermenter need to be mixed constantly to keep the fungus growing. However, the fungal hyphae of *Fusarium* are delicate, and a mechanical stirrer would damage them. Instead, a stream of compressed air provides more gentle agitation that protects the fungal hyphae.

The growth process in the fermenter lasts for six weeks. There is a steady input of nutrients into the fermenter during this time to maintain the growth of the fungus. The fungus converts the nutrients into biomass. Starch from potatoes or cereals is used as a source of carbohydrate because it is cheap and so makes the process economical. The starch is treated with enzymes to break it down to glucose before it is added to the fermenter. Ammonia is also added, as a source of nitrogen. The product is the whole fungal body of *Fusarium*, or its biomass.

Making food from mycoprotein

After being harvested from the fermenter, the mycoprotein is dried. At this stage it looks similar to pastry but at the microscopic level the harvested hyphae have a similar shape to animal muscle cells. Like muscle, they are made up of microscopic filaments. The fibrous texture of the mycoprotein makes it slightly chewy, similar to meat or fish.

Flavoured, shaped and packaged mycoprotein.

Science skills

a Use the data shown in the bar chart to make a table showing the nutrients in mycoprotein and their weight per 250 g of freshly harvested product.

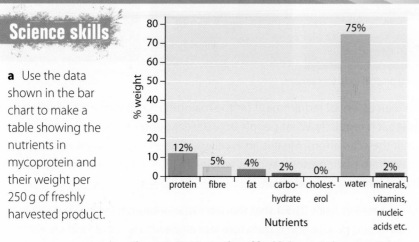

Figure 2 Nutrient value of freshly harvested mycoprotein.

Science skills

Table 1 shows the composition of chicken pieces, beefburgers, and some Quorn® products (food made from mycoprotein).

b Evaluate the health issues associated with each of the four products.

c Explain the difference in carbohydrate values for the two animal products.

d Give the percentage of saturated fat in the total oil/fat for each food.

Table 1 Composition of Quorn® and other foods.

Per 100 g	Quorn® pieces	Chicken pieces	Quorn® burgers	Beef burgers
energy/kJ	355.0	621.0	490.0	1192.0
protein/g	12.3	24.8	12.8	15.0
carbohydrate/g	1.8	0.0	5.8	3.5
oil/fat/g	3.2	5.4	4.6	23.8
of which saturates/g	0.6	1.6	2.3	10.0
fibre/g	4.8	0.0	4.1	0.4
sodium/g	0.2	0.1	0.5	0.5

Questions

1 **(a)** Which fungus produces mycoprotein? **(b)** What conditions does the fungus need to grow and reproduce?

2 Give two reasons why compressed air is pumped into the loop fermenter where *Fusarium* is grown.

3 **(a)** Use Figure 1 to explain how heat from the aerobic fermentation is prevented from overheating the fermenter so its temperature is a constant 32 °C. **(b)** Explain what might happen to enzymes if the temperature reached 45 °C.

4 **(a)** Give *two* nutrients that are added to the fermenter. **(b)** Explain why they are added throughout the fermentation process.

5 Explain precisely what the product from the fermentation is.

6 A vegetarian finds it difficult to include enough protein in his diet. Write a fact sheet for him to explain what mycoprotein is, an outline of its production, and three ways that it could be included in his diet.

Improving the efficiency of food production

Where does our protein come from?

Producing protein food for humans.

Protein is an essential requirement for the maintenance and growth of cells, so we must eat it regularly. What people actually eat depends on where they live and what they can afford. Meat is an excellent protein source, but it is more expensive than plant proteins such as whole grains. The reason is connected with the flow of biomass in food chains.

You will remember from B1 5.1 that there is a loss of biomass at every stage in a food chain. Food production is therefore less efficient when the food we eat is higher up the food chain.

Energy flow in food chains

Plants can only trap a tiny percentage (0.023%) of the energy in the sunlight for photosynthesis. At each stage in a food chain, energy is lost to the atmosphere as heat via respiration. At least 50% of the energy that primary consumers take in is lost as faeces. Animals that move around to search for prey use much of the energy from their food in the muscles, via respiration.

The units for energy flow are kilojoules per square metre per year (kJ/m²/year).

The data for producers show that of the 7 106 000 kJ/m²/year of light energy absorbed, only 87 000 kJ/m²/year was used in photosynthesis.

Of this 87 000 kJ (see Figure 1):

- 50 000 was used by the plants in respiration
- 14 000 was consumed by primary consumers
- 23 000 became detritus when producers died.

Food at lower levels in the food chain will have been produced more efficiently in terms of the energy it has retained.

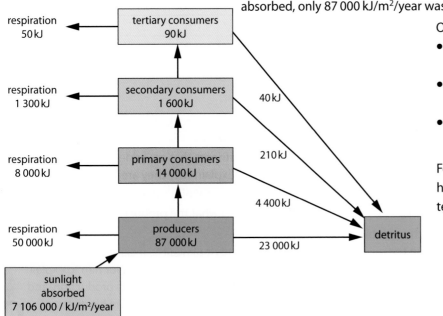

Figure 1 Energy flow in a stream.

a Tabulate the information in Figure 2.

b Display the information in another format.

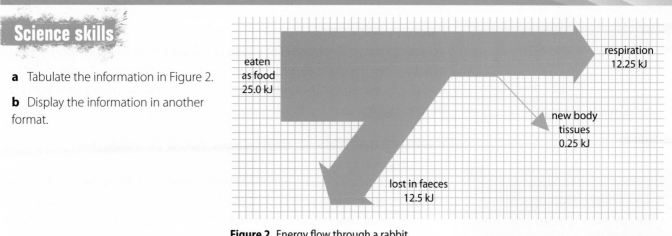

Figure 2 Energy flow through a rabbit.

Cereals or cows?

Energy that is wasted in the food chain does not get into our food. The higher our food is in the food chain, the less efficient the energy transfers to it. By using organisms for food that are low in the food chain, we can improve the efficiency of food production.

Figure 3 shows how many people can be fed from four hectares of land: an area about the size of five football pitches. The land will support 58 more people if they get their protein from soya beans than if their protein comes from beef.

10 acres (5 football pitches) will support:

60 people growing soya   10 people growing maize

24 people growing wheat   2 people rearing cattle

Figure 3 The number of people an area of land can support depends on how it is used.

1 (a) Explain why all animals need protein in their diets. (b) Give three different ways that protein from the environment can be harvested.

2 Suggest why it is less expensive to buy a 500 g loaf than it is to buy 500 g of minced beef.

3 Give a reason why predators use a higher proportion of respiratory energy in their muscles, than their prey use in their muscles.

4 Refer to Figure 1. (a) Calculate the proportion of energy absorbed by the producers that was used in photosynthesis. (b) Calculate the proportion of energy in the producers that was passed on to the primary consumers.

5 In which type of organism was the proportion of energy lost via respiration the highest?

6 Explain what is meant by detritus and how the biomass and energy in it are used.

7 Soya plants are about one-third of the height of maize plants and can be grown more densely than maize plants. Suggest why growing soya on a fixed area of land will feed six times more people than growing maize on the same area of land.

8 Describe energy flows along a food chain and, using this information, suggest how food production can be made more efficient.

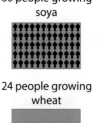

Intensive animal farming

- describe how energy losses from food animals can be restricted
- evaluate the positive and negative effects of managing food production by factory farming.

Which is the better way of rearing chickens?

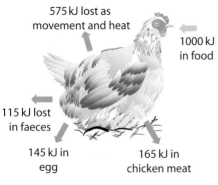

Figure 1 Energy flow through a free-range chicken.

575 kJ lost as movement and heat

1000 kJ in food

115 kJ lost in faeces

145 kJ in egg

165 kJ in chicken meat

Factory farming

At factory farms animals are kept indoors permanently in large numbers. Meat, milk and eggs are the main products of this industry. The aim of factory farming is to produce the most product at the lowest cost. There are risks and benefits to factory farming.

Rearing animals = inefficient food production

Many people like to eat meat, and factory farming supplies it as efficiently as possible. Even so, when you eat plant foods, all the material and energy from the plant goes directly into your body. If the same plant food was fed to an animal, and then the animal was killed and eaten, not all the original material and energy in the plant would be passed to your body. This is why rearing animals is not a very efficient way of producing food. Up to half of the world's harvest is fed to farm animals, while 800 million people go hungry.

Energy flow through mammals and birds

Cattle are primary consumers; they eat grass. The proportion of energy lost via respiration is high in cattle. This is because cattle are mammals. Mammals such as cattle, pigs, sheep and humans maintain a constant body temperature by generating heat inside the body. Keeping the body temperature high and constant needs a lot of energy from food.

Birds keep an even higher constant body temperature than mammals, so chickens and turkeys use an even higher proportion of their food in maintaining a constant body temperature.

Free-range or battery chickens?

Free-range chickens live outdoors for most of the time, where they feed mainly on natural food. Figure 1 shows how much of the energy we supply in food to a free-range chicken is transferred to human food.

Many hens that produce eggs are kept in cages, one above the other. This is battery hen farming. Keeping the hens in cages restricts their movement and so reduces energy loss.

Broiler chickens are reared for meat. Most of them are kept in poultry sheds holding up to 50 000 birds. They are not in cages. Poultry sheds are windowless buildings and have to be lit using artificial light.

Keeping hens in warm sheds reduces the amount of energy from food that they need to use to keep warm. Even so, almost half of the cereal food fed to the hens is not converted into meat or eggs. For factory-farmed chickens it takes 3.6 kg of cereal food to produce 2.0 kg of body mass.

It is generally acknowledged that rearing free-range chickens compared with factory farmed chickens provides the birds with a better standard of animal welfare. There is less disease and injury to the birds. Some consumers prefer to buy free-range poultry products for these reasons and some consumers also think the eggs and chickens taste better. However, free-range poultry products cost more to buy.

Farmers of free-range poultry have less expenditure on sheds and cages for the birds but they need more land to house a large flock of birds. Farming costs are reduced as there are no energy bills for heating and lighting the poultry houses.

a How much increase in egg production is there projected to be in developing countries from 1967 to 2015?

b What percentage of world egg production was in industrial countries between 1967 and 69?

c What percentage of world egg production is projected to come from developing countries in 2030?

Table 1 World egg production, past and projected.

Region	Million tonnes				
	1967–69	1987–89	1997–99	2015	2030
world	18.7	37.5	51.7	70.5	90.0
developing countries	4.9	16.2	33.7	50.7	69.0
industrial countries	10.7	12.8	13.7	14.8	15.5
transition countries	3.1	6.5	4.3	5.0	5.5

Zero-grazing dairy systems

Intensively farmed cows can produce around 10 times their natural milk yield. Choosing certain varieties of cattle and feeding them high-energy and protein food such as grain or food pellets helps produce this higher milk yield. However, this increase in production comes at a cost.

Cattle confined in sheds or outdoor yards stand permanently on concrete floors that are often covered with manure. The cattle often end up lame. Vast quantities of animal waste are produced from factory farms, and this is an important source of greenhouse gas emissions. Water may also be polluted by run-off from manure storage. The food concentrates given to the cattle can also cause problems. The outbreak of BSE in UK cattle in 1984 was linked to food pellets produced from diseased cattle and sheep.

What price cheaper food?

Questions

1   Explain what is meant by 'factory farms', and their aim.

2   Why is rearing animals for food inefficient?

3   Explain how in intensive poultry farms the movement of chickens is reduced and why it makes food production more efficient.

4   Poultry and cattle are kept in warm sheds on factory farms. Explain one advantage and one disadvantage of this for the farmer.

5   Why are there concerns about the spread of infectious diseases in intensive poultry farms?

6   **(a)** A farmer is deciding whether or not to keep battery hens. Make a table to show the advantages and disadvantages to him of farming battery hens. **(b)** Suggest other factors the farmer should think about before making a decision.

7   List two issues involved in factory farming that are of concern to you. Explain your choices.

8   Many people pay extra to buy free-range eggs. Outline an investigation to find if people can taste the difference between eggs from free-range hens and eggs from battery hens. Your investigation should be designed to give reliable results. You should include the independent variable, controls and how to measure the dependant variable.

Fish stocks

Is the ocean's supply of fish unlimited?

High-protein food from the sea

For thousands of years people living along coasts have harvested fish from the sea. But in recent years, over-harvesting has led to a collapse of fish stocks in many parts of the ocean.

The modern fishing industry can harvest fish extremely efficiently. Sonar can be used to pinpoint whole shoals of fish accurately. Enormous nets or lines are dragged through the oceans, sometimes for many hours. Large factory ships have on-board facilities for processing, packing and freezing fish.

Some spectacular collapses of fisheries have occurred because of overfishing. In 1992 the cod fishery off Newfoundland, Canada collapsed because the population dropped below that needed for sustainable breeding. Cod stocks in the North Sea, to the east of the UK, are also close to collapse.

Science in action

The Convention on International Trade in Endangered Species (CITES) can give protection to fish on the brink of commercial extinction; the bluefin tuna is one such fish. However, at a United Nations conference in Doha in March 2010, governments failed to pass a vote to protect the bluefin tuna. Some countries that voted against protection did so for economic reasons because their fishing fleets catch a high proportion of bluefin tuna. In other countries bluefin tuna is eaten as a delicacy so these consumers did not want the numbers caught to be reduced. Conflicting interests are often involved when trying to conserve an endangered species.

The bluefin tuna is on the brink of extinction.

Science skills

The Food and Agriculture Organization of the United Nations (the FAO) publishes a two-yearly report on the state of the world's fisheries. Fish stocks that are fully exploited are in danger of over-exploitation, at which stage they collapse.

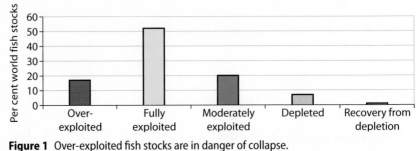

Figure 1 Over-exploited fish stocks are in danger of collapse.

a What percentage of world fish stocks have not been affected by fishing?

b What percentage of world fish stocks are neither over-exploited nor fully exploited?

Using food webs to manage fish stocks

Scientists believe that the complex food web that exists in the ocean must be taken into account if fisheries management is to be successful. **Ecosystem overfishing** is when the balance of the ecosystem is altered due to overfishing. For example, the abundance of large predatory fish such as swordfish, tuna, cod and halibut has declined sharply. Ninety per cent of stocks of these fish have already gone. As a consequence, there is an increase in abundance of their prey. These are small foraging fish such as gurnard, wolf fish and black bream.

Sustainable fishing

Sustainable fishing means catching a consistent amount of fish over an indefinite period without otherwise damaging the environment. This is the aim of fisheries management organisations. Unless they are fished sustainably, certain species may disappear altogether in some areas. Mature adult populations must be able to produce enough offspring to keep the population stable. Cod reach maturity at about 3–4 years of age, when they are about 50 cm long. The minimum length of cod that can be caught under EU regulations is 35 cm, which is too small for sustainability.

Practical measures that can be taken to maintain fish stocks include:

- controlling net size
- fishing quotas.

A net's mesh size determines what size fish the net will catch. Fish that are smaller than the mesh can pass through. If the mesh size is right, juvenile and immature fish can escape and live to reproduce. EU laws govern the minimum mesh size for specific target fish.

The Common Fisheries Policy (CFP) of the EU sets limits or quotas for each country on the amounts of fish they can catch. Each fishing vessel is given an individual fishing quota for different fish types. These compromises for the fishing industry are imposed to create a basis for a viable fishing industry in the future.

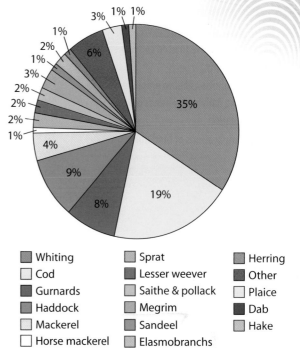

Figure 2 Food that fish in waters around the UK prefer to feed on.

Whiting, Cod, Gurnards, Haddock, Mackerel, Horse mackerel, Sprat, Lesser weever, Saithe & pollack, Megrim, Sandeel, Elasmobranchs, Herring, Other, Plaice, Dab, Hake

Mesh size controls which fish are caught.

Questions

1. What has caused the collapse of fish stocks in the oceans?
2. Describe what factory ships do.
3. Why do modern fishing vessels use sonar?
4. **(a)** Name a species of fish on the brink of commercial extinction. **(b)** Suggest why the species you mention has not been given protection.
5. Why must ocean food webs be considered in order to manage commercial fisheries successfully?
6. What is the aim of fisheries management organisations?
7. Controlling net size is a conservation measure to protect fish stocks. Explain how it works.
8. Explain what the competing priorities are that require a compromise in the form of fishing quotas.

Feeding the world

Agriculture, water and food security

The United Nations Food and Agriculture Organization (FAO) published the following statistics relating to the global supply of food and water.

- The daily drinking water requirement per person is 2–4 litres, but it takes 2000–5000 litres of water to produce one person's daily food.
- It takes 1000–3000 litres of water to produce one kilogram of rice and 13 000–15 000 litres to produce one kilogram of grain-fed beef.
- Over the period 2010–2050, the world's water supply will have to support agriculture to feed an extra 2.7 billion people.

A billion people in the world today are malnourished, while in more developed countries over-consumption is creating a global crisis of obesity. The aim of the FAO is to achieve food security for every household. A household is considered food secure when its occupants do not live in hunger or fear of starvation.

Science skills

a Tabulate and display the data on the map in graphical form for the following countries: UK, South Africa, India, Australia, Chile and USA.

b Suggest two reasons for the uneven consumption of food worldwide

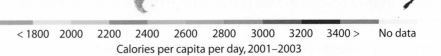

| < 1800 | 2000 | 2200 | 2400 | 2600 | 2800 | 3000 | 3200 | 3400 > | No data |

Calories per capita per day, 2001–2003

Figure 1 World food consumption 2001–2003.

Solutions to reduce global hunger

The United Nations World Food Programme (WFP) provides food assistance to save lives in disasters and emergency situations. In 2010, an earthquake devastated Haiti and there were severe floods in Pakistan. In Haiti, part of the emergency response consisted of distributing food aid with the goal of feeding children first. WFP also has training and education programmes to help poor people and communities become self-reliant and produce their own food.

The long-term solution to food security must involve the growth of more **staple crops**. These are crops such as cereals, the seeds of which can be kept all year round to provide energy and protein. This may require a conversion from growth of vegetables like carrots or leafy vegetables to cereals. It may also require more land for crop use in developed countries, for example amenity or national park land.

One way to increase yields of staple food may be to use genetically modified or GM crops. To date more than 39 million hectares of GM crops have been grown

GM crops are widespread in America. In Europe many countries have banned their growth.

worldwide. Scientists have concerns about the potential effects of GM crops on human health, the environment and biological diversity.

Are the world's taps running dry?

The world's six billion people are using 54 per cent of all the accessible freshwater contained in rivers, lakes and underground **aquifers**. This is putting great pressure on the natural systems that recycle and replenish water supplies. The world cannot increase its supply of fresh water, but we can change the way we use it. For instance, management of water in agriculture should be improved. About 20 per cent of all crop land is irrigated. This irrigation accounts for 60 per cent of world water use. Drip irrigation and low-pressure sprinklers use far less water than conventional irrigation. If these techniques were widely adopted this would reduce water waste.

Irrigation needs careful management.

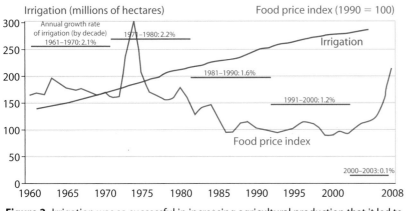

Figure 2 Irrigation was so successful in increasing agricultural production that it led to a 30-year decline in food prices.

Figure 3 The World Food Programme logo.

Are food miles fair miles?

Developed countries import food from around the world to fill the supermarket shelves. **Food miles** are the distance food is transported from the place it is grown until it reaches the consumer. Some people have expressed concern about the carbon emissions produced when food is transported large distances. They argue that consumers in developed nations should avoid buying green beans from Peru or winter strawberries from Kenya.

However, the economy of less economically developed nations often relies heavily on food exports. If consumers boycott their goods, the livelihoods of farmers in those countries may be threatened.

Questions

1 Explain what is meant by food security.

2 Give an example of where the WFP has supplied food aid and the types of food supplied.

3 How much more water is needed to produce one kilogram of beef than one kilogram of rice?

4 **(a)** How may GM crops be advantageous to world food supply? **(b)** Give two concerns that some scientists have about the potential effects of GM crops.

5 **(a)** How is water used extensively in agriculture and how could its use be better managed? **(b)** Look

at Figure 2. **(i)** What was the increase in irrigation between 1970 and 2000? **(ii)** What was the difference in the price index between 1970 and 2000?

6 List three natural types of fresh water stores on Earth.

7 Evaluate the implications of using the term 'food miles' to make consumers aware of energy used to transport food we buy.

8 Discuss the options available to increase the world supply of staple crops by 2050.

ISA practice: how does temperature affect fermentation?

A bread company has found a new yeast, and wants to investigate the fermentation of the yeast at different temperatures.

Section 1

1 Write a hypothesis about the fermentation at different temperatures for a yeast. Use information from your knowledge of fermentation to explain why you made this hypothesis. *(3 marks)*

2 Describe how you are going to do your investigation.

   You should include:
   - the equipment that you are going to use
   - how you will use the equipment
   - the measurements that you are going to make
   - a risk assessment
   - how you will make it a fair test.

   You may include a labelled diagram to help you to explain your method.

   In this question you will be assessed on using good English, organising information clearly and using specialist terms where appropriate. *(9 marks)*

3 Design a table that you could use to record all the data you would obtain during the planned investigation. *(2 marks)*

   Total for Section 1: 14 marks

Section 2

Study Group 1 was a group of students who carried out an investigation into the hypothesis. They used several different temperatures of yeast and sugar solution. Figure 1 shows their results.

At 18°C the volume of CO_2 produced was 12 cm^3
At 23°C the volume of CO_2 produced was 16 cm^3
At 28°C the volume of CO_2 produced was 21 cm^3
At 33°C the volume of CO_2 produced was 25.5 cm^3
At 38°C the volume of CO_2 produced was 30 cm^3
At 43°C the volume of CO_2 produced was 35 cm^3

Figure 1 Study Group 1's results.

4 **(a)** Plot a graph of these results. *(4 marks)*

   (b) What conclusion can you draw from the investigation about a link between temperature and the rate of fermentation? You should use any pattern that you can see in the results to support your conclusion. *(3 marks)*

   (c) Do the results support the hypothesis you put forward in answer to question 1? Explain your answer. You should quote some figures from the data in your explanation. *(3 marks)*

Below are the results from three more study groups.

Study Group 2 was a second group of students who carried out a similar investigation. They used different temperatures. Figure 2 shows their results.

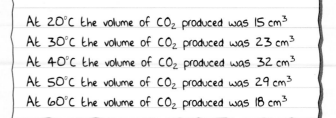

At 20°C the volume of CO_2 produced was 15 cm^3
At 30°C the volume of CO_2 produced was 23 cm^3
At 40°C the volume of CO_2 produced was 32 cm^3
At 50°C the volume of CO_2 produced was 29 cm^3
At 60°C the volume of CO_2 produced was 18 cm^3

Figure 2 Study Group 2's results.

Study Group 3 was a third group of students. They decided that they wanted to know what happened throughout the experiment, not just at the end. Table 1 shows their results.

Table 1 Results from Study Group 3.

Time after first gas released/min	Volume of gas collected/cm^3		
	18 °C	22 °C	26 °C
0	0	0	0
2	1	2	3
4	2	5	12
6	4	13	20
8	7	20	28
10	10	28	36
12	17	35	41

Study Group 4 were a group of scientists working for the bread company. They measured the loss in mass of the yeast and sugar solution at different temperatures. Instead of using just 100 cm³ of sugar solution they used 1000 cm³, and left the experiment for 1 hour. Their results are shown in Table 2, and plotted as a graph in Figure 3.

Table 2 Study Group 4's results.

Time / min	Loss in mass/grams			
	20 °C	30 °C	40 °C	45 °C
0	0	0	0	0
10	21	32	45	34
20	42	66	90	59
30	64	94	130	75
40	84	119	136	76
50	102	136	137	77
60	118	137	138	78

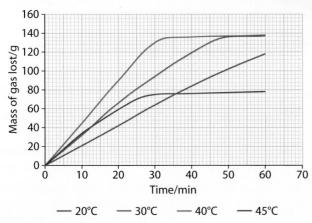

— 20°C   — 30°C   — 40°C   — 45°C

Figure 3 Graph of results from Study Group 4.

5 **(a)** Draw a sketch graph of the results from Study Group 2. *(3 marks)*

   (b) Look at the results from Study Groups 2 and 3. Does the data support the conclusion you reached about the investigation in question 5(b)? Give reasons for your answer. *(3 marks)*

   (c) The data contain only a limited amount of information. What other information or data would you need in order to be more certain whether or not the hypothesis is correct?
   Explain the reason for your answer. *(3 marks)*

   (d) Look at Study Group 4's results. Compare them with the data from Study Group 1. Explain how far the data support or do not support your answer to question 5(b). You should use examples from Study Group 4's results and from Study Group 1. *(3 marks)*

6 **(a)** Compare the results of Study Group 1 with Study Group 2. Do you think that the results for Study Group 1 are *reproducible*?
   Explain the reason for your answer. *(3 marks)*

   (b) Explain how Study Group 1 could use results from other groups in the class to obtain a more *accurate* answer. *(3 marks)*

7 Suggest how ideas from the original investigation and the other studies could be used by the bread company to decide the best temperature at which to ferment the bread (let it rise) and the optimum time for fermentation. *(3 marks)*

Total for Section 2: 31 marks
Total for the ISA: 45 marks

Assess yourself questions

1 (a) Give two reasons why tropical rainforests are being cut down at a rapid rate. *(2 marks)*

(b) Cutting down rainforests reduces biodiversity.

Give one consequence for humans of a reduction in the biodiversity of tropical forests. *(2 marks)*

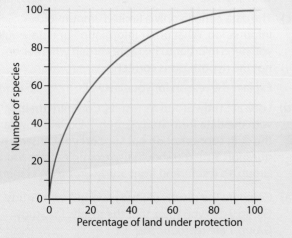

Figure 1 Graph of number of species preserved versus percentage of land under protection.

(c) Biodiversity can be preserved by protecting parts of rainforests. Refer to Figure 1.

(i) Describe, the relationship between the percentage of land protected and number of species preserved. *(2 marks)*

(ii) Explain why small-scale protection projects can be very effective. *(2 marks)*

2 (a) Explain what is meant by 'sustainable development'. *(2 marks)*

(b) China has the largest population in the world. Demand for electricity there is increasing rapidly. Suggest two reasons for this increase in demand. *(2 marks)*

(c) Table 1 shows how China plans to change its methods of generating electricity by 2020.

Year	Percentage of electricity obtained from energy source					
	hydroelectricity	coal	oil	gas	nuclear	other
2000	24.8	69.3	4.8	0.3	0.7	0.1
2020	27.1	58.6	1.6	7.5	4.2	1.0

Table 1 Changes in electricity generation in China.

Will the changes to China's electricity production help the environment? Use data from the table to explain your answer. *(3 marks)*

3 An 'ecological footprint' is the measure of how much land and water a human population needs to produce the resources required to sustain itself and to absorb its wastes.

• The average American uses 25 hectares to support his or her current lifestyle.

• The average Canadian uses 18 hectares.

• The average Italian uses 10 hectares.

(a) Suggest why Americans have a larger ecological footprint than Italians. *(2 marks)*

(b) Suggest three ways in which ecological footprints can be reduced. *(3 marks)*

4 Table 2 gives the energy output from some agricultural food chains.

Food chain	Energy available to humans from food chain/kJ per hectars of crop
cereal crop ⟶ humans	800 000
cereal crop ⟶ pigs ⟶ humans	90 000

Table 2 Energy output from two farm food chains.

(a) Explain why the food chain:
cereal crop ⟶ humans
gives far more available energy than the food chain:
cereal crop ⟶ pigs ⟶ humans. *(4 marks)*

(b) Explain, in terms of energy, why many farm animals are kept indoors in temperature-controlled conditions. *(4 marks)*

(c) Read the passage below about a pig farm in the USA.

Inside barns, long rows of grunting, snorting pigs fill every available space. Each row contains 100 animals, all pregnant or soon to be. Every animal faces the same direction in a scene of orderliness seldom associated with pigs.

The animals are not lining up by choice; each stands inside a narrow metal crate. The pigs, which can reach 270 kg, will spend much of their three or four years of adult life inside these crates, unable to turn around or even lie down fully because the stalls are just two feet wide. Only when caring for piglets will the sows live outside them for long, and then in different metal crates only slightly wider so they can recline to nurse. This farm outside Chicago is by all accounts a model of pork industry efficiency, cleanliness and productivity, and the metal 'gestation crates' are nothing unusual in the nation's highly industrialised pork business.

But critics of this kind of intensive pig farming, people ranging from animal welfare activists to academic researchers and some big pork buyers, have been raising increasingly pointed and sometimes emotional objections to the crates. Some call the practice inherently cruel, some call it offensive because the confinement produces abnormal behaviours in relatively intelligent animals, and some worry it could endanger the pork industry if consumers begin to focus on it.

Mainstream pork buyers are beginning to take note of the farm animal welfare issue. McDonald's, with its finely tuned understanding of consumers, especially the young, has assembled a task force of outside experts on animal welfare and production specialists to study whether pork suppliers should be required to find alternatives to sow crates.

Pig farmers believe they are being unfairly accused of ignoring the wellbeing of their animals. Industry officials say the individual stalls contribute to animal welfare by ensuring that all the sows are fed, and by preventing otherwise frequent fighting among animals. What's more, they say, crates keep prices low by allowing many sows to be raised, watched and controlled in relatively small areas, and they improve food safety by keeping animals cleaner.

(i) Give three advantages to farmers of keeping sows in gestation crates. (3 marks)

(ii) Give three groups of people who are opposed to gestation crates. (3 marks)

(iii) Give two reasons why some people are opposed to the use of gestation crates. (2 marks)

(iv) Why might firms such as McDonald's put pressure on farmers to ban gestation crates? (2 marks)

5 Figure 2 shows how mycoprotein is produced.

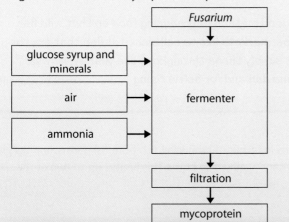

Figure 2 Flowchart of mycoprotein production.

(a) What type of organism is *Fusarium*? (2 marks)

(b) Name two types of nutrients that are used in this process. (2 marks)

(c) Explain why ammonia is necessary for this process. (2 marks)

(d) Sales of foods made from mycoprotein are increasing rapidly. Suggest an explanation for this. (2 marks)

6 Figure 3 shows a biogas generator on a farm in a developing country.

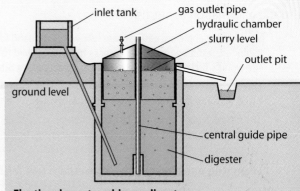

Floating dome-type biogas digester

Figure 3 Biogas generator.

(a) Suggest what is fed into the inlet tank. (2 marks)

(b) Which gas is the main constituent of biogas? (2 marks)

(c) Suggest the function of the floating gas dome. (2 marks)

(d) Suggest why the material in the outlet pit is useful to the farmer. (2 marks)

(e) Figure 4 shows how temperature affects the volume of biogas produced by a biogas generator.

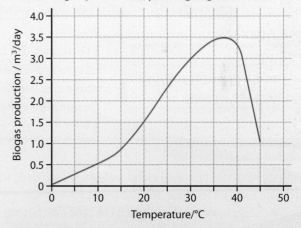

Figure 4 Effect of temperature on volume of biogas.

(i) Describe the effect of temperature on biogas production. (3 marks)

(ii) Explain the effect of increasing the temperature above 38 °C on biogas production. (2 marks)

(iii) The mean winter temperature in this developing country is 15 °C. The farmer uses the biogas mainly for cooking. Explain one disadvantage of this type of biogas generator for the farmer. (2 marks)

Here are three students' answers to the following long-answer 6-mark question.

Read the information about fish farming.

> Worldwide people are eating more fish. As the demand for fish rises, populations of both marine and freshwater species are being over-exploited, resulting in declining stocks from many wild fisheries. This has resulted in fish being farmed on a large scale, providing 42% of the world's fish supply. More space on land and at sea is being taken up every year by fish farms.
>
> Fish farming predictably delivers uniform-sized fillets of fish all year round, meeting the demands of global supermarkets. Farmed fish often sells at a lower cost than wild fish as the farms have become more productive.
>
> The future expansion of fish farming, however, may be jeopardised by several looming issues. For instance, there are growing concerns about coastal pollution from poorly run fish farms, in the form of excess feed and manure. There is a higher incidence of disease in farmed fish compared with wild fish. Escaped fish carrying disease originating on farms can devastate wild fisheries.
>
> Using the information above, evaluate fish farming as a method of managing food production.

Read the answers together with the examiner comments. Then check what you have learnt and try putting it into practice in any further questions you answer.

In this question you will be assessed on using good English, organising information clearly and using specialist terms where appropriate.

B Grade answer

Student 1

The answer should not start with a conclusion.

No credit is given for copying text from the passage.

Do not use words such as 'good' or 'bad' because they imply a moral judgement.

> I think that fish farming is a good method of producing food so that wild fish stocks are not depleted. Supermarkets like to sell pieces of fish that are the same size and fish farms can supply these throughout the year. Worldwide people are eating more fish and demand for fish is rising.

Examiner comment

In questions of this type, two marks are awarded for 'pros', two marks are awarded for 'cons' and two marks for a conclusion. The conclusion mark is not awarded if the candidate has not given both 'pros' and 'cons'. This candidate has given three reasons in favour of producing fish from fish farms, but has given no reasons against fish farms. This limits him to two of the six marks.

Read the question carefully.
- When answering evaluation questions first read the passage. Mark up 'pros' and 'cons' as you go through.
- These questions carry a maximum of 6 marks.
- The first paragraph of your answer should contain at least two 'pros'.
- Write a second paragraph including at least two 'cons'.
- Write a conclusion that includes your opinion about the issue you are evaluating and the word 'because', so that you give a reason for your conclusion.

 **Grade answer**

Student 2

It is much better to keep 'pros' and 'cons' separate rather than alternating them.

It should be made clear whether this is a 'pro' or a 'con'

No marks are given for a conclusion that does not give a reason.

Fish farming increases the world food supply without depleting wild fish stocks by over-fishing. Wild fish stocks may be threatened though, if fish carrying diseases escape from fish farms and spread into natural populations. To supply 42 per cent of the world's fish, more space on land and at sea is taken up by fish farms. In the supermarkets farmed fish sells at a lower price than unfarmed fish. I think that the case for fish farming is stronger than the case for reducing the number of fish farms.

Examiner comment

The answer has 'pros' and 'cons' alternating with each other, but the candidate does give two reasons in favour of fish farming: it does not deplete fish stocks and it produces fish that sell at a lower price than wild fish. Two reasons against fish farming are also given: the fact that diseased fish may escape and threaten wild fish stocks, and that more space on land and in the sea is taken up annually by fish farms. Although the candidate expressed an opinion about the value of fish farms she did not follow this up with the word 'because' and a reason, so she did not get the conclusion marks. This answer would be awarded 4 marks.

A* **Grade answer**

Student 3

The candidate makes it quite clear that he is talking about the 'pros'.

The candidate makes it quite clear that he is moving onto the 'cons'.

The candidate now starts a third paragraph for the conclusion and uses the word 'because' to explain the reason for his decision.

The main case for fish farming is that it supplies fish on a large scale without depleting wild fish stocks, which can suffer population crashes if over-fished. Fish farming can deliver a reliable fish supply of uniform quality throughout the year.

The case against fish farming is that environmental pollution may result from manure and fish food released into the water. In addition diseased fish may escape from fish farms and cause infection in wild fish stocks, reducing their number.

Overall I think that fish farming should be encouraged because it supplies high-protein food more cheaply and reliably then catching wild fish, and it could help to secure the human food supply for the next decade. Tighter controls and licensing of fish farms would help to solve the problems of pollution that fish farms may cause.

Examiner comment

The answer clearly directs the examiner to two 'pros', two 'cons' and a reasoned conclusion. The conclusion refers to both a 'pro' and a 'con' and then goes on to suggest how the 'con' may be reduced in the future. Six marks would be awarded.

Examination-style questions

1   The diagram below shows what happens to the shape of a plant cell placed in distilled water.

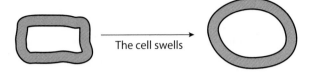

The cell swells

(a) Explain the changes to the cell that you can see in the diagram.   (2 marks)

(b) Describe the change that will occur if a piece of peeled potato is placed in a concentrated sugar
     solution, and explain why this change occurs.   (3 marks)

2   Some scientists investigated the amounts of sulfate ions taken up by barley roots in the presence of oxygen
    and when no oxygen was present. The graph below shows the results.

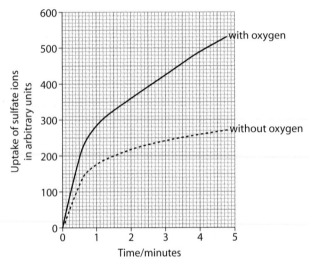

(a) Calculate the difference between the rate of uptake of sulfate ions with and without oxygen between
     1 and 2 minutes. Show clearly how you work out your answer.   (3 marks)

(b) The barley roots were able to take up more sulfate ions with oxygen than without oxygen. Explain how. (3 marks)

3   The diagram below shows the structure of a villus.

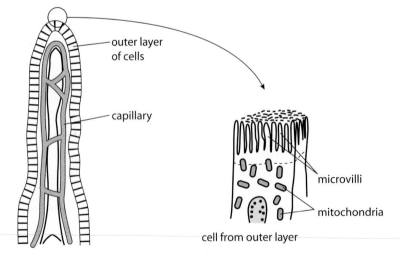

Using information from the diagram, explain how the villus is adapted for absorption of the soluble
products of digestion.   (4 marks)

4   The table below gives information about a geranium plant and a cactus plant. The geranium grows in gardens in the UK. The cactus grows in hot deserts.

Feature	Geranium	Cactus
thickness of waxy cuticle/μm	5	15
total leaf surface area/cm^2	1800	150
percentage of water storage tissue in stem	50	85
number of stomata/mm^2	59	13
time of day when stomata open	daylight	at night
horizontal spread of roots/m	0.2	5

Using only information in the table, explain how the cactus is better adapted for living in hot, dry conditions.

In this question you will be assessed on using good English, organising information clearly and using specialist terms where appropriate.   (6 marks)

5   The graph below shows blood pressure measurements for a person at rest. The blood pressure was measured in an artery and in a vein.

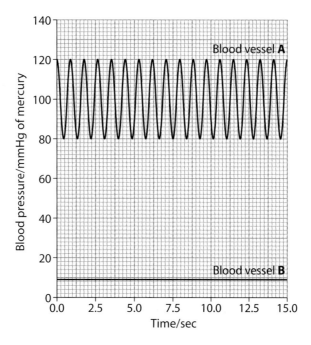

(a) Which blood vessel, **A** or **B**, is the artery? Give *two* reasons for your answer.   (2 marks)

(b) Calculate the person's heart rate per minute.   (2 marks)

(c) Describe the way in which each of the following substances is transported to the tissues. You should include the site at which each substance enters the blood.

   (i) Glucose.   (3 marks)

   (ii) Oxygen.   (3 marks)

6 (a) Explain fully why glucose is found in the blood, but not in the urine. *(3 marks)*

(b) The table below shows the concentrations of dissolved substances in the urine of a healthy person and the urine of a person with one type of kidney disease.

Substance	Concentration / g/dm³	
	Urine of healthy person	Urine of person with kidney disease
protein	0	6
glucose	0	0
amino acids	0	0
urea	21	21
mineral ions	19	19

Suggest an explanation for the difference in composition of the urine between the healthy person and the person with the kidney disease. *(2 marks)*

7 The graph below shows the effect of different internal body temperatures on a person's rate of energy loss by sweating.

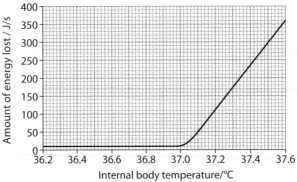

(a) How much more energy was lost from the body each second by sweating when the body temperature was 37.6 °C than when it was 36.6 °C? Show clearly how you work out your final answer. *(2 marks)*

(b) Explain why a person would feel more thirsty when the body temperature was 37.6 °C than when it was 36.6 °C. *(2 marks)*

(c) Explain how sweating helps to control body temperature. *(3 marks)*

8 Cod in the North Sea have been overfished and the stock of fish is depleted. The graph below shows the biomass of fish capable of reproducing, for the years 1967 to 2007.

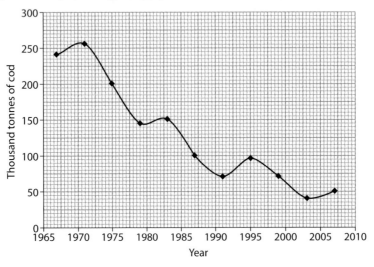

(a) What is meant by sustainable fishing? *(2 marks)*

(b) Give two conservation measures that have been taken to protect fish stocks. *(2 marks)*

(c) 150 000 tonnes of cod biomass is described as the minimum biologically acceptable level for fishing. What does the graph show in relation to this? *(2 marks)*

9 The human population is increasing rapidly and human activities produce waste.

(a) Give three parts of the environment that can be polluted if waste is not properly handled, and for each give a possible pollutant. *(3 marks)*

(b) List four ways that humans cause less space to be available for wildlife. *(2 marks)*

10 Large-scale deforestation in the tropics can provide land for agriculture to produce more food. Scientists are concerned that food production has led to changes in the amount of methane in the atmosphere. The graph below shows the methane concentration in the atmosphere from 1989 to 2009.

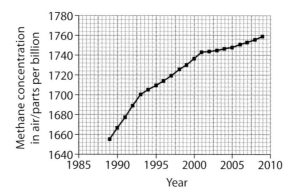

(a) (i) Describe the trend in the data. *(2 marks)*

(ii) Give one plant and one animal crop that produce methane. *(2 marks)*

(iii) Explain why scientists are concerned about the increase of methane in the atmosphere. *(3 marks)*

(b) The photograph below shows a palm oil plantation in tropical Malaysia where harvesting is taking place.

(i) Suggest which part of the palm oil tree is used for the extraction of palm oil for use as a biofuel. *(1 mark)*

(ii) Evaluate oil palm cultivation on deforested land in the tropics for the production of biofuel. *(6 marks)*

In this question you will be assessed on using good English, organising information clearly and using specialist terms where appropriate.

Glossary

A

2,4-D – (2,4-dichlorophenoxyacetic acid) A common selective weedkiller, used against broadleaf plants.

abdomen – The area under the ribs, between the diaphragm and the pelvis.

accuracy – The closeness of a measurement to the true value.

active transport – The transport of molecules against a difffusion gradient using energy from respiration.

adaptation – A characteristic that helps an organism to survive in its environment.

adult cell cloning – A process in which the nucleus of an adult cell is inserted into an unfertilised egg cell from which the nucleus has been removed, and the cell divides and develops into a new individual.

adult stem cell – A stem cell extracted from differentiated tissue, which can normally differentiate only into a limited range of specialised cells.

aerobic respiration – Respiration that requires the presence of oxygen to release energy from glucose, producing carbon dioxide and water.

agar – A nutrient-rich jelly obtained from seaweed used for growing microbes.

alcohol – In drinks, ethanol, obtained by fermenting fruit or grain. A legal recreational drug.

allele – One form of a gene. Different alleles of the same gene produce slightly different characteristics, such as different eye colours.

allergic reaction – A reaction of the body to a compound to which the immune system has been sensitised.

alveolus (plural alveoli) – A tiny air sac in the lungs. The alveoli have a large total surface area over which oxygen diffuses into the blood and carbon dioxide diffuses out.

amino acid – A simple compound which, when combined with other amino acids in chains, makes proteins. There are 20 types of amino acids commonly found in living cells.

amylase – A type of digestive enzyme that catalyses the breakdown of starch into glucose.

anaerobic digestion – The breakdown of dead plant and animal material without oxygen.

anaerobic fermentation – Breakdown of organic matter by bacterial action without oxygen.

anaerobic respiration – Respiration which doesn't need oxygen – the release of energy from glucose without oxygen. In muscle cells this produces lactic acid as a waste product. The process releases less energy than aerobic respiration.

antibiotic – A chemical compound, often produced by microbes, that destroys other bacteria.

antibody – A chemical compound produced by white blood cells when the immune system detects the presence of a pathogen. Antibodies are specific to a particular pathogen, that is, each type of antibody attacks only one type of pathogen.

antigen – A cell marker protein on the outside of a cell. Antigens differ between people.

antiseptic – A chemical compound used to stop the growth of microorganisms on living tissue.

antitoxin – A chemical compound produced in the body by white blood cells to neutralise poisons produced by microbes.

antiviral – A drug used to treat infections caused by viruses, such as flu or measles.

aorta – The principal artery leading from the heart.

aquatic – Living in water.

aquifer – An underground area of rock holding groundwater that can be extracted for use by humans.

artery – A blood vessel carrying blood away from the heart (usually oxygenated blood).

asexual reproduction – Reproduction from one cell of a parent organism by simple cell division, without fertilisation of gametes, that is, without fusion of male and female sex cells.

atrium – A chamber in which blood enters the heart. The left atrium receives blood from the lungs, the right atrium receives blood returning from the rest of the body.

autism – A disorder in development in which a person finds it difficult to communicate with others and to interact socially.

autoclave – A device used for sterilising laboratory equipment with superheated steam.

autotrophic – Able to produce its own food from simple starting compounds. Plants are autotrophic, because they can photosynthesise.

auxin – A plant hormone that affects growth, rooting and flower and fruit formation.

B

bacterium – A single-celled microorganism without a nucleus. Bacteria can reproduce very quickly, and some cause disease.

balanced diet – A diet made up of a variety of different foods to provide the energy and all the nutrients needed to stay healthy.

bile – A mixture of chemicals produced in the pancreas that emulsifies fats, and so helps digest them.

binding site – The site in an enzyme to which the substrate binds.

biomass – Biological material; in particular, the total mass of living material at a specified level in a food chain or in a specified area.

bone marrow – Fatty tissue inside long bones that contains adult stem cells capable of producing all types of blood cell.

brain – The centre of the nervous system, responsible for interpreting sensory signals, co-ordinating motor responses and higher activities such as thought and memory.

breathing rate – The number of breaths taken in a measured time, for example per minute.

by-product – A substance produced in a chemical process that is not the main product being made.

C

caffeine – A drug found especially in coffee and tea that is a mild stimulant to the central nervous system.

callus – A cluster of cells grown by tissue culture.

camouflage – Body colouring and/or shape that helps conceal an organism in its environment.

cannabis – An illegal drug that can produce a sense of euphoria (well-being), but also anxiety and loss of short-term memory. Also called marijuana.

capillary – A very small blood vessel of about the width of a red blood cell, with walls one cell thick to allow substances to move in and out; capillaries penetrate almost all tissues and transport substances to and from cells.

carbohydrase – An enzyme that catalyses the breakdown of starch to sugars.

carbohydrate – A food type that is the main energy source in the diet. Carbohydrates, for example sugars and starch, contain only the elements carbon, hydrogen and oxygen.

carbon cycle – The way in which carbon atoms circulate between living organisms and the physical environment.

carbon store – A place in which carbon is kept out of the atmosphere for many years, for example a tree.

carrier – A person carrying one allele for a recessive disorder, so they do not have the disorder themselves but could pass it on to their children.

cast – A fossil formed when the space left by the decay of a dead organism fills with minerals that harden into the shape of the organism.

catalyst – A chemical compound that speeds up a reaction but is not itself used up.

categoric variable – A variable that can only take particular values, for example days of the week or different types of food.

cell – The basic unit of all known organisms. Most cells have a nucleus, a cell membrane and other organelles with their own functions.

cell membrane – A thin outer layer of a cell that controls what goes into and comes out of a cell.

cell sap – A liquid that contains dissolved sugars and salts and is found in plant cell vacuoles.

cell wall – The outermost layer of a plant cell. It provides shape and support for the cell.

cellulose – A carbohydrate that forms tiny fibres from which plants and algae build cell walls.

central nervous system – The brain and spinal cord.

cerebral palsy – A medical condition in which the brain does not co-ordinate momvement properly. It is caused by damage to the brain before birth or in early childhood.

characteristic – A feature of an organism, such as its size, colour or behaviour; many characteristics are coded for by genes.

chemical energy – Energy stored in the chemical bonds of compounds, such as those in the cells of living organisms.

chemosynthesis – The construction of molecules, such as glucose, from smaller molecules as building blocks.

chlorinated – With chlorine added to kill bacteria and other microorganisms.

chlorophyll – A green pigment found in the chloroplasts of plant cells. It captures light energy for photosynthesis.

chloroplast – An organelle in plant cells that contains chlorophyll for photosynthesis.

cholesterol – A fatty substance made in the liver and used in cell membranes.

chromosome – An immensely long molecule of DNA containing many regions called genes, each of which carries the genetic information that influences a characteristic of the organism. Chromosomes are found in the nuclei of cells.

circulatory system – A series of blood vessels through which blood flows around the body.

classification – A system for grouping organisms according to their characteristics.

climate change – Alterations in the long-term climate. The climate change we can observe at present is known to be caused by the increased concentration of CO_2 in the atmosphere caused by human activity.

clinical trial – A set of tests on people to provide data on the effectiveness of a drug.

clomiphine – A fertility drug that works by blocking the effect of oestrogen on the pituitary gland.

clone – An individual that is genetically identical to the parent because it is produced by mitosis.

cocaine – An illegal, highly addictive stimulant drug produced from the leaves of the coca plant (*Erythroxylum coca*). It produces feelings of euphoria (well-being).

combustion – Burning; the reaction that occurs when a fuel combines with oxygen in a heat-producing oxidation reaction.

common ancestor – An organism or species that is a shared ancestor (in evolutionary terms) of two or more later organisms or species.

competition – The struggle between individual organisms for a share of a limited resource, such as water or food; for example, competition between two predators hunting the same prey.

compost – Remains of plants and other organic waste broken down by microorganisms; it can be mixed with soil for use as a garden fertiliser.

compost heap – A pile of organic, usually garden and kitchen, waste that decays to produce compost.

concentration – A measure of how much solute is dissolved in a solvent such as water.

concentration gradient – A difference in the concentration of a substance in two different areas, for example two different solutions of the same solute on opposite sides of a partially permeable membrane.

constrict – Become narrower.

continuous variable – A variable that can take any value, such as the time taken for a chemical reaction to happen.

continuous variation – Variation within a population of animals or plants of a characteristic that can take any value and that is strongly affected by the environment, for example height.

control variable – A variable that must be kept constant throughout an experiment to ensure that it does not affect the dependent variable.

core body temperature – The temperature in the middle of the body, around the organs of the heart, lungs and liver, which is normally around 37.6°C.

cover – *See* ground cover.

criterion – Guidance on which decisions can be judged.

culture medium – A solution or jelly containing nutrients for the growth of microorganisms.

culture plate – A plate coated with a jelly containing nutrients for the growth of microorganisms.

cutting – A small piece of a plant, such as leaf or stem, used to grow a new plant.

cystic fibrosis – An inherited disorder caused by a recessive allele that results in production of thick, sticky mucus, affecting the lungs and other parts of the body.

cytoplasm – The substance outside the nuclei in cells, in which many of the chemical reactions take place. These reactions are catalysed by enzymes.

D

decay – The breakdown of dead plant and animal material by fungi, bacteria and other organisms.

deficiency disease – A disease caused by a lack of an essential nutrient in the diet.

dehydrated – A person or organism suffering from loss of water is said to be dehydrated.

denature – To alter the shape of an enzyme (a protein molecule) usually by heating it, in such a way that it no longer performs its function.

dependent variable – A dependent variable is the variable that is measured for each change in the independent variable in an experiment.

detritus feeder – An organism that feeds on detritus (decomposing organic waste).

diabetes – A disease in which blood glucose concentration is not controlled.

dialyser – A partially permeable membrane inside a machine through which blood is passed for haemodialysis. Substances diffuse across the membrane between the blood and a dialysing fluid, restoring their correct concentrations in the blood.

diaphragm – In mammals, a sheet of muscle across the bottom of the rib cage, which increases the volume of the thorax, and therefore also the volume of the lungs, when it contracts. This function is an important part of breathing.

differentiated – (Of a cell) Specialised to carry out a specific job. Muscle cells, for example, are specialised to be able to contract.

diffusion – Movement of a substance by random motion from a region of high concentration to a region of lower concentration.

digest – Break down into smaller particles, as when food is digested in the gut.

digester – A large vessel in which dead plant or animal material is broken down anaerobically by microorganisms, for example in in-vessel composting.

dilate – Become wider, increase in diameter; for example, a blood vessel or the pupil of the eye can dilate.

directly proportional – A directly proportional relationship between two variables is a simple mathematical relationship: if one variable is doubled, for example, the other is doubled too.

disinfectant – A substance used to kill microorganisms on non-living surfaces.

distribution – The area of the environment in which a species lives.

DNA (deoxyribonucleic acid) – The genetic material found in the nucleus of living cells. Chromosomes are made up of DNA, and a gene is a section of a chromosome.

DNA profiling (DNA fingerprinting) – A process that produces an image of variable sections of DNA, used to identify individuals. Previously called DNA fingerprinting.

dominant – In genetics, an allele is dominant if it produces its form of the characteristic in the organism even if only one chromosome of the pair carries that allele (that is, whether the organism is homozygous or heterozygous).

dose – The amount received at one time, for example a dose of insulin administered as medicine or of radiation in a CT scanner.

double helix – The shape of the long DNA molecule, with two strands twisting about one another.

double-blind trial – A clinical trial in which neither the subjects nor the experimenters know which substance is the test substance and which is the control.

drought – A period when there is a lack of water for organisms.

drug – Any substance that, when absorbed into the body, alters normal body function. Medical drugs, such as antibiotics, are used to fight illness or disease. Recreational drugs, such as alcohol and caffeine, can be depressants or stimulants, or give a sense of wellbeing.

E

ecological relationship – The relationship of an organism with other organisms and their physical environment in an ecosystem.

ecosystem overfishing – Ecosystem overfishing occurs when too many fish have been caught from a body of water, causing fish stocks to fall below a healthy level for the ecosystem.

ecstasy – The illegal synthetic drug MDMA (3,4-methylenedioxymethamphetamine).

ectothermic – Animals are ectothermic if they use external methods like basking in the sun or moving into the shade to control their body temperature. These animals, such as amphibians and reptiles, need to absorb heat from their surroundings to keep warm.

effector – A cell or an organ that brings about a response to a stimulus.

effluent – A waste liquid that flows out when a process is complete.

egest – Expel undigested waste material from the gut.

electron microscope – A microscope that uses electrons rather than light to create images. It has a much higher magnification than a light microscope.

embryo – A new individual in its first stage of development, when a fertilised egg cell starts to divide.

embryo screening – Testing an embryo to see if it carries certain alleles, such as those for a genetic disorder.

embryo transplanting – Splitting early embryos into separate cells and placing them into the wombs of other, surrogate, mothers to develop.

embryonic stem cell – A stem cell extracted from an embryo, which can differentiate into all the kinds of specialised cell found in that organism.

energy store – A deposit of starch that a plant can convert back into glucose when it needs more energy from respiration.

enzyme – A protein molecule that acts as biological catalyst to speed up the rate of a reaction taking place within or outside a cell.

enzyme technology – The use of enzymes as industrial catalysts.

epidemic – The widespread outbreak of an infectious disease within a country.

epithelial cell – A cell that is part of the outer layer (the epithelium) of a structure or an organ.

ethene (ethylene) – The smallest alkene, C_2H_4. Contains a double bond between the carbon atoms and is often used as a chemical feedstock.

ethical question – A question concerning what is morally right or wrong.

evolution – Changes in the characteristics of species over time.

evolutionary relationship – The existence of similar characteristics in different organisms because they evolved from a common ancestor.

evolutionary tree – A diagram that shows how a group of organisms evolved from earlier organisms.

excrete – Discard waste products from cell reactions within the body, such as urine from kidneys.

expiration – The movement of gases from the lungs out into the air.

extinct – (A species) having no individuals still living.

extracellular – Outside the cell.

extremophile – An organism that is adapted to an extreme condition of the environment.

F

faeces – Waste matter remaining in the gut after food has been digested, which is egested from the body.

family tree – Diagram that shows the members of a family linked by relationship, which can be used to show inheritance of a genetic disorder.

fat – An organic compound, not miscible with water, that forms the main part of cell membranes and is important for energy storage.

fatigue – Tire and lose level of response, as in muscles during extended vigorous activity.

feedback control – Method of controlling a system in which a change causes the opposite change to occur, to maintain a value within limits.

fermentation – One kind of anaerobic respiration by microorganisms.

fertilisation – The fusion of a male gamete with a female gamete to produce a cell with two sets of chromosomes (a zygote).

fertilised – Refers to an egg cell in which male gametes have joined with the cell nucleus to form a zygote (an egg cell that can develop into a new animal or plant).

fixing – Absorbing carbon dioxide from the air and converting it into complex carbon compounds. Plants fix carbon dioxide during photosynthesis.

flagellum (plural flagella) – A tiny 'tail' on a cell that enables the cell to move.

food chain – A diagram that shows who eats what in a single set of feeding relationships.

food miles – The distance food is transported from the place it is grown until it reaches the consumer.

food web – A diagram that shows all the feeding relationships between organisms living in the same area.

forensic science – Scientific methods used to help identify what happened in a crime, including using DNA profiling to identify people involved.

fossil – The remains of an organism that lived in the past found preserved in rock, or evidence of organisms having been there (such as footprints).

fossil fuel – Fuel formed millions of years of years ago from the remains of ancient animals and/or plants, such as coal, crude oil and natural gas.

FSH – The hormone that, in women, stimulates follicles in the ovaries to release an egg; follicle-stimulating hormone.

G

gall bladder – An organ that stores bile and releases it into the small intestine.

gamete – A specialised sex cell, such as sperm or eggs, involved in sexual reproduction in plants and animals.

gas chromatography – A method that separates chemicals in a very small sample. It can be used to separate fragments of DNA to make a DNA profile.

gateway drug – An illegal drug that does not have seriously harmful effects, but can lead users on to consumption of more harmful drugs.

gel – A jelly-like material with a grid structure that can trap water.

gel electrophoresis – A procedure that separates chemicals in a liquid mixture. It was used to make DNA profiles, but has been largely replaced by gas chromatography, which can analyse smaller samples.

gene – Small section of DNA in a chromosome. Each gene contains the code for a particular inherited characteristic, that is, to make a particular protein.

genetic – Relating to genes and DNA (the hereditary material).

genetic diagram – A diagram that shows the inheritance of alleles or characteristics from parents by offspring. Punnett squares are a type of genetic diagram.

genetic engineering – The process of taking genes from one organism and putting them into the cells of another so that the cells include the characteristic of the new gene.

genetically modified (GM) – An organism that has been genetically engineered.

gland – Any small organ responsible for the production and release of a particular substance for the control of bodily functions by chemical means.

global warming – The rise in mean surface temperatures on the Earth, thought to be due to increasing amounts of greenhouse gases such as carbon dioxide.

glomerulus – The capillary network in the top of a kidney tubule where unwanted substances are filtered from the blood.

glucagon – The hormone produced by the pancreas that causes cells in the liver to release glucose into the blood.

glucose – A simple sugar (carbohydrate) produced in plants by photosynthesis and from starch by digestion, broken down in respiration to release energy.

glycogen – A form of carbohydrate made from glucose in animals. It is stored in muscle and liver cells, then broken down when glucose levels in the blood are low, for example during vigorous exercise.

gravitropism – Growth (of a plant root) downwards.

greenhouse – A glass building, sometimes heated, that is used for growong plants outside their normal growing season.

greenhouse effect – The trapping of warmth by greenhouse gases such as carbon dioxide in the Earth's atmosphere that keeps the surface of the Earth warm enough for life.

ground cover – The area of ground covered by a particular plant species.

H

Haber process – An industrial process named after the German chemist Fritz Haber, used to make ammonia from nitrogen.

haemodialysis – A process in which blood from a patient with kidney failure is passed through a dialyser with an artificial membrane to remove urea and restore the normal balance of water, glucose and mineral ions.

haemoglobin – A protein that transports oxygen around the body in the blood and gives red blood cells their colour.

HBOC – A type of artificial blood, haemoglobin-based oxygen carrier.

heart rate – The number of heartbeats in a measured time, usually one minute.

herbicide – A chemical that kills plants, used to treat weeds in crops.

heroin – An addictive, illegal, recreational drug.

homeostasis – The processes that keep variations of some factors, such as temperature, water, mineral ions and blood sugar concentration, within limits in the body.

homeothermic – Animals that keep their body temperature roughly constant, regardless of the surrounding temperature.

hormone – A chemical produced in one part of the body and transported in the blood to control a process in another part of the body; insulin and glucagon are hormones.

host mother – Female organism that has had an embryo from another female implanted in her womb.

hydrothermal vent – A crack in the ocean floor where heat and chemicals escape from the rocks below into the water.

hydrotropism – Growth (of a plant root) towards water.

hygiene – Hand-washing, cleaning floors and surfaces, and other procedures designed to reduce the chances of infection from bacteria and other microorganisms.

hypha (plural hyphae) – A microscopic, narrow, branched, thread-like fungal structure.

hypothesis – An idea that is suggested to explain a set of observations. It is used to make predictions that can be tested scientifically.

I

immune – Protected against disease by the production of antibodies.

immune system – The organs and mechanisms that protect an organism against pathogens and disease.

immunisation – Administration of a vaccine to provide immunity against a disease or diseases.

immunosuppressant drug – A drug that suppresses the reaction of the immune system, used to prevent rejection of transplanted organs such as kidneys.

impulse – The form in which information is transmitted by nerve cells.

in vitro fertilisation (IVF) – The fertilisation of an egg in the laboratory by the addition of sperm.

incubation – Gentle warming to grow microbes or hatch eggs in an incubator.

incubator – A container in which the temperature and humidity are controlled, used to hatch eggs or grow microorganisms.

independent variable – A variable that is changed or selected by the investigator.

ingest – Take food into the body.

inherit – In biology, to receive a genetic characteristic from a parent.

inoculating loop – A small loop of wire with a handle, used to spread a microbial culture on an agar plate.

inoculation – Injection with microbes (living or dead).

insoluble – Unable to dissolve in a particular solvent, usually water.

inspiration – Breathing in.

insulate – Act as a barrier to the transfer of energy by heating, or to the conduction of electricity.

insulin – A hormone produced by the pancreas that controls the concentration of glucose in the blood.

insulin pump – Machine for delivering insulin continuously just below the skin.

intercostal muscle – One of the muscles between the ribs that move the ribs up and out in breathing.

in-vessel composting – Forming compost from waste plant material within a large vessel in which conditions of temperature and moisture can be controlled.

ion – An electrically charged particle, containing different numbers of protons and electrons. An ion is an atom or molecule that has either lost (positively charged) or gained (negatively charged) one or more electrons.

ionising radiation – High-energy radiation that ionises atoms or molecules by removing electrons. Ionising radiation includes some ultraviolet radiation, X-rays and gamma rays.

iron lung – The first type of negative-pressure ventilator.

IVF (*in vitro* fertilisation) – A procedure carried out in the laboratory involving mixing eggs and sperm removed from the reproductive organs to encourage fertilisation.

K

kidney – An organ in the body that maintains water and mineral ion balance in the blood and removes waste substances such as urea from the blood.

kidney failure – The condition in which kidneys function at less than 30% of their normal level.

L

lactic acid – A breakdown product of anaerobic respiration in muscle cells.

leprosy (Hansen's disease) – A serious infectious disease of the skin and nerves caused by *Mycobacterium* bacteria.

LH – The hormone, produced by the anterior pituitary gland, that triggers ovulation.

lichen – An organism formed from a fungus and lichen that can live in extreme conditions; used as an indicator of air pollution.

light intensity – The amount of light energy falling on a measured surface area, for example one square metre.

limiting factor – An environmental variable, such as light intensity, that limits the rate of a process, such as a chemical reaction.

lipase – An enzyme that catalyses the digestion of lipids (fats or oils) into fatty acids and glycerol.

lipid – A fat or oil.

liver – An organ that produces bile and many other compounds necessary for life.

M

malnutrition – The result of eating an unbalanced diet. Malnutrition can cause illness or make people prone to catch diseases.

mass extinction – An event in which a large number of species become extinct during the same period.

maturation – The process of becoming an adult.

mean – The arithmetical average of a set of data.

median – The middle value of a set of values arranged in number order. The median of the values 3, 4, 4, 6, 7, 7, 9 is 6.

medicine – The science of preventing, alleviating or curing disease, illness or injury.

meiosis – A type of cell division to produce gametes. Two divisions of the original cell produce four cells with half the normal number of chromosomes.

menstrual cycle – The monthly cycle of changes in a woman's reproductive system, controlled by hormones.

metabolic rate – A measure of the energy used by an animal in a given time period.

methanogen – A microorganism that produces methane.

microbe – A microorganism.

microorganism – An organism that is too small to be seen without a microscope.

mineral ion – Small charged particle, such as a sodium ion or a chloride ion, that must be balanced in cells for healthy functioning. Animals obtain minerals from food; plants absorb them from the soil.

mitochondrion (plural mitochondria) – An organelle (structure) within a cell in which aerobic respiration takes place.

mitosis – A form of cell division that produces two cells genetically identical to the parent cell.

MMR – A vaccination for measles, mumps and rubella (German measles).

mode – The value that occurs most often among a set of data.

monitor – Continually check for changes.

motor neurone – A nerve cell that carries information to an effector organ, such as a skeletal muscle.

mould – The shape formed by a dead organism pressing into soft sediment.

moult – Shed fur, feathers, skin or scales. For example, cold-adapted mammals lose some of their thick fur in summer.

MRSA – A type of bacteria resistant to antibiotics (methicillin-resistant *Staphylococcus aureus*).

mummy – A type of fossil in which soft tissue is well preserved.

mutation – A change in a gene that may result in different characteristics.

N

natural selection – A natural process whereby the organisms with genetic charateristics best suited to their environment survive to reproduce and pass on their genes to the next generation.

negative linear – A straight-line plot in which one variable decreases in proportion to the increase in the other variable is a negative linear plot.

negatively gravitropic – Tending to grow away from the pull of gravity.

negative-pressure ventilator – A ventilator placed around the body that causes air to be sucked into the lungs.

nervous system – The body system that transmits sensory information and muscle signals, co-ordinates body action, and is reponsible for thinking and memory. It consists of the brain, spinal cord and neurones or nerves.

net movement – The overall movement of particles from an area of higher concentration to one of lower concentration.

neurone – A cell that transmits electrical nerve impulses, carrying information from one part of the body to another.

nicotine – The active ingredient of the recreational drug tobacco; a stimulant that causes dependence.

nitrate – A compound containing ions with the formula NO_3^-.

nucleus – The large, membrane-bound organelle inside a cell that contains genetic material.

nutrient – A substance that a living thing needs so that it can live healthily.

O

obese – A condition in which excess body fat has accumulated to the extent that it may have an adverse effect on health.

oestrogen – A sex hormone produced in the ovary that is responsible for egg release and for female secondary sexual characteristics.

organ – A body structure that has a specific function and is made up of several different types of tissue.

organelle – A structure found within a cell that has a particular function, such as the nucleus or a chloroplast.

organic – Relating to or derived from living organisms.

osmosis – The movement of water molecules from a region of higher water concentration to a region of lower water concentration through a partially permeable membrane.

ovary – Female reproductive organ in humans and other animals. Ovaries produce eggs.

ovum – An egg cell.

oxygen debt – The extra oxygen that the body needs after vigorous exercise.

oxyhaemoglobin – The bright red protein in arterial blood that is a combination of haemoglobin with oxygen from the lungs. It transports oxygen to the cells of the body.

P

palisade mesophyll cell – A leaf cell packed with chloroplasts. Most photosynthesis takes place in palisade mesophyll cells.

pancreas – An organ that produces digestive enzymes and controls blood sugar concentration by producing insulin.

pandemic – A disease that is spread rapidly across many countries.

partially permeable membrane – A thin membrane containing tiny pores that allow the smallest particles to pass through, but not others.

pathogen – A microorganism that causes illness or disease.

penicillin – One of a group of antibiotics produced from *Penicillium* fungi.

permeable – Allowing the passage of water or a particular material.

pesticide – A chemical used to kill pests such as insects or crop diseases.

Petri dish – A shallow, round glass or plastic dish with a lid; used for growing microorganisms.

PFC – A type of artificial blood containing polyfluorocarbons, which is entirely synthetic.

pharmaceutical – A drug used for medical purposes.

phloem – A plant tissue made up of sieve tubes and companion cells that transports sugar.

photosynthesis – The process by which green plant cells produce sugars and oxygen out of carbon dioxide from the air, water from the soil and energy from sunlight.

phototropism – Growth towards or away from the light.

physical condition – A condition of the environment caused by physical processes, such as temperature and amount of water.

physiotherapy – Treatment for a problem or disorder that involves physical activity, such as firm patting on the back for cystic fibrosis suffers to loosen sticky mucus.

placebo – A dummy drug that looks identical to the real thing, given to people in the control group in clinical trials to ensure that the people receiving the drug and the control group are treated exactly the same in all other ways.

placebo effect – An improvement in a condition produced by taking a placebo although it contains no active ingredient.

plankton – Microscopic single-celled organisms found in lakes, rivers and the sea.

plasma – The liquid part of the blood.

plasmid – A loop of genetic material outside the main chromosome in a bacterium.

plasmolysis – Shrinkage of cell cytoplasm caused by placing the cell in a solution more concentrated than the cytoplasm so that it loses water by osmosis.

platelet – A small fragment of a cell in the blood, which plays a part in producing a clot at a wound site.

poliomyelitis – A disease caused by a virus that can paralyse its sufferers; it has been almost eradicated by vaccination. Also called polio.

pollination – The transfer of pollen from one flower to another so that fertilisation is possible.

pollution – Damage to the environment or to living things caused by the careless release of waste.

pollution indicator – Organisms whose presence or absence indicates the existence of pollution in a given area.

polydactyly – An inherited condition that causes a person to have more fingers, thumbs or toes than usual, in some cases caused by a dominant allele.

polypeptide – A long-chain molecule made by joining many amino acids together. Proteins are long polypeptides.

polytunnel – A long arched structure of poly(ethene) sheet used as a greenhouse in agriculture.

positive linear – A straight-line plot in which one variable increases in proportion to the other variable is a positive linear plot.

positively gravitropic – Tending to grow toward the pull of gravity.

positively hydrotropic – Tending to grow toward sources of water.

positively phototropic – Tending to grow toward the light.

positive-pressure ventilator – A ventilator that forces air into the lungs.

potometer – An apparatus for measuring the rate at which water is taken up by a plant.

predation – The killing and eating of one kind of organism by another kind of organism, usually in terms of animal predators and their prey.

producer – Organisms that make their own food, e.g. plants and green algae by photosynthesis, or some bacteria by chemosynthesis.

product – A compound formed during a chemical reaction.

protease – A digestive enzyme that catalyses the breakdown proteins into amino acids.

protein – A large molecule made of amino acids linked together; proteins play important roles in all living things and are an important part of the human diet.

pulmonary artery – The artery through which deoxygenated blood is pumped from the heart to the lungs.

pulmonary vein – The vein through which oxygenated blood from the lungs returns to the heart.

Punnett square – A table used to predict the probability that a characteristic will be inherited by the offspring of two organisms with known combinations of alleles. One form of genetic diagram.

pure culture – A culture containing only one type of microorganism.

pure-breeding – Homozygous (and therefore certain to pass a particular allele on to offspring so that all the offspring of two pure-breeding organisms have that characteristic only).

pyramid of biomass – A diagram that shows the mass of living organisms at each stage in a food chain.

Q

quadrat – A frame used for sampling the distribution of species in an area.

R

radioactive dating – Measuring the ratios of radioactive elements in rock to work out when the rock formed.

radioactive isotope – An unstable nucleus, which at some point will decay into a different nucleus with a different number of protons and neutrons, emitting radiation when it does so.

random error (uncertainty) – An unpredictable variation around the true value, causing each reading to be slightly different.

random sampling – Sampling by a scheme generated randomly rather than according to a pattern.

randomised controlled trial – A way of making the test of a new drug or treatment fair and ensuring that the expectations of patients or doctors do not skew the results. Patients are randomly put into either a treatment group or a control group, and neither the doctor nor the patient knows which group any patient is in until the trial is over.

range – The spread between maximum and minimum values in a set of experimental results.

receptor – An organ or a cell that is sensitive to external stimuli.

recessive – In genetics, an allele is recessive if it produces its form of the characteristic in the organism only if both chromosomes of the pair carry that allele (that is, the organism is homozygous).

recreational drug – A drug that is taken for personal satisfaction rather than for medical purposes.

rehydrate – Replace the water that has been lost or used up.

rejection – Rejection occurs when the immune system attacks and destroys a transplant organ.

relay neurone – A neurone that carries information from a sensory nerve cell to a motor nerve cell.

resistant – A bacterium that has antibiotic resistance is less susceptible to the action of antibiotics than other bacteria.

respiration – The breakdown of glucose in cells to release energy, carbon dioxide and water.

resuscitator – A hand-controlled ventilator most often used by paramedics.

ribcage – The protective structure formed by the ribs.

ribosome – Tiny organelles inside a cell where protein synthesis occurs.

root hair cell – A cell in the root of a plant with long narrow projections that extend into the soil to absorb water and dissolved mineral ions.

root tip squash – The preparation of a root tip by gently squashing it, so that the individual cells can be seen clearly under a microscope.

rooting powder – A powder containing the plant hormone auxin, which encourages a cutting from a plant to grow roots.

rotting – The decomposition of biological material by microorganisms.

runner – Horizontal stems that some plants produce in order to create new plants by asexual reproduction.

S

salivary gland – One of several glands in the mouth that produce the digestive enzymes amylases.

sampling – Examining a small portion of an area and using the results to estimate the value for the total area.

selective weedkiller – One of a group of chemicals that kill broad-leaved weeds but not wheat or other grasses.

sensory neurone – A nerve cell that carries information to the brain or spinal cord.

sequestration – Carbon sequestration is the removal of carbon dioxide from the atmosphere and storage where it cannot add to climate change.

sex chromosome – A chromosome that determines the sex of the individual: human females have two X chromosomes, human males have one X and one Y chromosome.

sexual reproduction – Production of offspring by the fusion of male and female sex cells (gametes).

shivering – The action of small, rapid contractions by the muscles to generate heat if core body temperature gets too low.

side effect – An unwanted reaction to a drug or other medicine; for example, some people are allergic to penicillin.

simvastatin – A type of drug called a statin, designed to reduce levels of cholesterol in the blood.

small intestine – A section of the digestive tract that produces protease, amylase and lipase enzymes.

soluble – Able to dissolve in a particular solvent, usually water.

solute – A substance that is dissolved in a liquid to make a solution.

specialised cell – A cell that has special features that are required for its function. For example retina cells are specialised to sense light. Also called differentiated cells.

speciation – The evolution of new species from one original species.

species – A group of organisms that have many characteristics in common and are able to breed together.

spinal cord – The cord running down the centre of the spine, an important part of the central nervous system, which carries sensory and motor nerve impulses to and from the brain.

stable – Not easily changed.

staple – A staple food is one that forms the main part of the diet. Staple foods are high in carbohydrate, for example wheat, potatoes and maize.

staple crops – Crops with a high carbohydrate content that can provide staple foods.

starch – The form of carbohydrate stored by most plants. Starch is made from glucose molecules joined together.

statin – A drug that reduces levels of cholesterol in the blood.

stem cell – A cell that, unlike most body cells, can divide and differentiate into other cell types.

sterilised – Treated to kill all microbes.

steroid – A member of a group of chemicals that include hormones such as oestrogen and medical drugs. Some athletes take anabolic steroids to build muscle.

stoma (plural: stomata) – A tiny hole in the surface of leaves for the exchange of gases between the air spaces inside the leaf and the air outside.

stomach – An organ in the digestive tract that produces protease enzymes and acid.

storage organ – A special underground organ, such as a potato, in which plants store food over winter.

substrate – The molecule upon which an enzyme acts.

succession – The gradual, predictable process by which the species in a habitat change over time until a stable, 'climax' ecosystem is produced.

symptom – A sign of a disease or disorder noticed by the patient, such as high temperature and runny nose.

synapse – The junction of two nerve cells.

systematic sampling – Sampling according to a pattern rather than randomly.

T

territory – An area defended by an animal against other animals.

testes – The male reproductive organs in humans and other animals. The testes produce sperm.

testosterone – A male sex hormone, produced in the testes. It is responsible for the development of the male reproductive organs and other male characteristics.

test-tube baby – A baby produced by *in vitro* fertilisation (IVF).

thalidomide – A sedative drug introduced in the 1960s. It was found to cause serious deformities in infants when taken by pregnant mothers.

theory of evolution – The theory that explains how all living things today have been produced by the accumulation of inherited changes in the characteristics of populations through successive generations.

thermoregulatory centre – The part of the brain that monitors and controls the core body temperature.

thorax – The section of the body between the neck and the ribs.

tissue – A mass of similar cells, for example muscle tissue.

tissue culture – Growth of cells and/or tissues outside the animal or plant. Whole plants can be grown from plant cells using tissue culture.

tissue type – The type of tissue defined by the cell marker proteins on the cells in the tissue. These markers vary from person to person.

toxic – Poisonous.

toxicity – A measure of how poisonous a drug or other compound is.

toxin – A poison (often one produced by microbes).

trace fossil – A fossil that provides evidence of the presence of organisms, such as footprints, burrows or root shapes, rather than a fossil of the organism itself.

transect – A line across an area along which the species are sampled in a field study.

transgenic organism – An organism that contains genes from another organism.

transpiration – The loss of water through the leaves of plants.

transplant – Incorporating a healthy kidney from a donor into a patient with kidney failure.

trophic level – A level in a food chain, such as producer or primary consumer.

tubule – A fine tube in the kidneys where waste and useful substances are exchanged between the filtrate inside the tubule and a blood capillary beside it.

turgor – The pressure on a cell wall caused by the water in the cell. Turgid plant cells help give rigidity to the plant.

type I diabetes – Diabetes caused by the inability of the pancreas to produce insulin, meaning that glucose is not removed from the bloodstream when its level rises. Without the injection of insulin, the person would eventually have so much glucose in the blood that they would pass into a coma and die.

U

ultraviolet – Electromagnetic radiation that has a wavelength shorter than visible light but longer than X-rays.

unicellular organism – An organism that has only one cell, such as a bacterium.

unidirectional – In one particular direction only.

urea – A waste product formed in the liver from the breakdown of excess amino acids that the body does not need, and excreted from the blood through the kidneys.

urine – A waste liquid produced in the kidney tubules after the exchange of waste and useful substances between the liquid and the blood.

uterus – The womb, the part of the female reproductive system where a fetus develops before birth.

V

vaccination – The administration of a vaccine in order to stimulate the body to develop immunity to a particular pathogen.

vaccine – A preparation of dead or inactive pathogens administered to prompt white blood cells to make antibodies to destroy live pathogens of that type; often injected.

vacuole – A sac filled with cell sap found in most plant cells.

valve – A mechanism that permits flow in one direction only; for example, the valves in the heart that prevent blood flowing in the wrong direction.

variegated – Of more than one colour; a variegated leaf has green areas containing chlorophyll and yellow/white areas with no chlorophyll.

vascular bundle – Xylem and phloem tissues for transport in plants.

vein – A blood vessel carrying deoxygenated blood returning to the heart. Veins are thinner than arteries and contain valves.

vena cava – A large vein that returns deoxygenated blood into the right atrium of the heart.

ventilation – The movement of air in and out of the lungs.

ventricle – A chamber in the heart from which blood is pumped into the arteries. The left ventricle pumps blood around the body, the right ventricle pumps it to the lungs.

villus (plural villi) – A tiny, finger-like projection from the inner surface of the intestine. Villi increase the surface area of the intestine, increasing the absorption of nutrients.

virus – A microbe much smaller than a bacterium, which reproduces inside cells, where it is not susceptible to antibiotics. Viruses are the cause of measles, flu and the common cold.

vitamin – An organic compound needed in the diet in small quantities for good health.

W

weed – A plant that grows where it is not wanted.

white blood cell – A cell that circulates in the blood and is an important part of the immune system. White blood cells are the body's main line of defence against disease.

windrow composting – Outdoor composting on a large scale.

X

xylem – A plant tissue made up of long hollow tubes that act as vessels for the transport of water.

Y

yield – The amount of a crop that a plant produces.

Z

zygote – The cell formed when two gametes fuse to create a new individual in sexual reproduction.

Index

2,4-D (2,4-dichlorophenoxyacetic acid) 30

A

abdomen 190
acid rain 60
active transport 184, 187
adaptations of animals 52–53
adaptations of plants 50–51
adult stem cells 156
aerobic respiration 57, 146–147
agar gel 18
Agent Orange 30
agriculture 250
alcohol 36
algae 110
alleles 158–159
alleles, dominant 160
alleles, recessive 160
allergic reactions 144
alveoli 107, 188–189
amino acids 136, 206
amylase 140
anaerobic digestion 67
anaerobic fermentation, biogas 236–237
anaerobic fermentation, ethanol biofuels 234
anaerobic respiration 57, 150–151
animal adaptations 52–53
animal cells 100–101
animal farming, intensive 246–247
antibiotics 12
antibiotics, resistance to 14
antibodies 10, 213
antigens 213
antiseptics 9
antitoxins 10
antiviral drugs 15
aorta 194
aquatic organisms 61
aquifers 251
arteries 194, 196
artificial aids to breathing 190–191
artificial blood 199
artificial hearts 201
asexual reproduction 74–75, 155
athletics, steroids in 40–41

B

atrium 194
autism 17
autoclave 18
auxins (plant hormones) 28
auxins (plant hormones), uses 30–31

bacteria 9, 104
bacteria, chemosynthetic 56
bacteria, killing 12
balance of substances in organisms 206–207
balanced diet 4
bees, effect of changing environments 59
bile 141
binding sites 136
biodiversity 124
biodiversity, reduction in 230
biofuels 230
biofuels, containing ethanol 234–235
biogas, collection from landfill and lagoons 240–241
biogas, large-scale production 238–239
biogas, small-scale production 236–237
biological detergents 144
biomass and energy 62–63
biomass pyramid 62
birds, effect of changing environments 59
blood 198–199
blood, artificial 199
blood, cholesterol level 5
blood, circulation 194–195, 196
blood, controlling glucose levels 216–217, 218
blood, transportation of oxygen 199
blood, transportation of substances 198
blood, white cells 10–11, 198
blood vessels 196–197
body temperature 22–23, 214–215
bone marrow 156
brain 20
breathing in humans 190–191
breathing in humans, artificial aids 190–191
breathing rate 148

C

caffeine 37
callus 76
camouflage 54
cannabis 37
cannabis, mental illness 38–39

capillaries 194, 197
capillaries, constriction of 214
capillaries, dilation of 214
carbohydrases 142
carbohydrates 4, 64
carbon compounds in animals 68
carbon cycle 68–69
carbon dioxide capture 68
carbon stores 230, 233
carriers of disorders 163
casts 166
catalysts, biological 137
categoric variation 72
cell division 154–155
cell membrane 20
cell membranes 100
cell sap 102
cell walls 102
cells, animal 100–101
cells, bacterial 104
cells, differentiation of 108
cells, differentiation of 156–157
cells, diffusions of gases 106
cells, diffusions of liquids 106
cells, plant 102–103
cells, specialised 100–101
cells, yeast 105
cellulose 102
cellulose, synthesis of 115
central nervous system (CNS) 36
cerebral palsy 27
characteristics of organisms 72
cheating in sport 40–41
chemical energy 62
chemosynthetic bacteria 56
chickens 246
chlorinated handwash 9
chlorophyll 102
chloroplasts 102, 110
cholesterol 5
chromatography, gas 159
chromosome sets 154
chromosomes 72, 158
chromosomes, sex 160
circulation of the blood 194–195, 196
circulatory system 194, 196
classification of living things 82
climate change 232
clinical trials 34
clomiphine 27

clones 74, 155

cloning 76–77

cloning, adult cells 77

cocaine 37

cold, animal adaptations to 52

cold, plant adaptations to 50

combustion of fuels 68

common ancestors 83

communities of organisms 122–123

companion cells 202

competing for resources 54

competition 168

compost 231

compost heaps 67

concentration gradient 106, 184

concentration of substances 106

constriction of capillaries 214

continuous variation 72

contraceptive pill 25

contraction of muscles 20

core body temperature 214

coronary heart disease 200

crops, environment manipulation 118–119

crops, staple 250

crops, yield 116

culture medium 18

culture plates 19

cuttings of plants 75

cystic fibrosis 163, 164

cytoplasm 20

cytoplasm 100

D

dairy systems, zero-grazing 247

data collection, environmental 233

data, analysing 126–127

data, collecting 124–125

decay of dead waste 64–65

deficiency diseases 4

deforestation 230–231

denaturing of proteins 138

detergents, biological 144

detritus feeders 68

diabetes 4, 216

diabetes, treating 218–219

diabetes, types 217

dialyser 210

dialysis 210–211

diaphragm 190

diet 4–5

dieting (slimming) 6–7

differentiation of cells 108, 156–157

diffusion 184, 188

diffusion of gases 106

diffusion of liquids 106

diffusion through cell membranes 107

digesters 67, 238–239

digesting pathogens 10

digestion 216

digestion, anaerobic 67

digestive enzymes 140

dilation of capillaries 214

diseases, defences against 10–11

diseases, spreading 9

diseases, treating and preventing 12–13

disinfectants 9

distribution of species 59

DNA 100, 158

DNA fingerprinting 159

DNA profiling 159

dominant disorders 162

dominate alleles 160

dose of insulin required 219

double circulation 195

double helix of DNA 158

double-blind trials 34

drought 50

drug, definition 36

drug, gateway 38

drugs, developing 34–35

drugs, failure of testing 35

drugs, recreational 36–37

drugs, recreational, harm caused 38–39

dry conditions, animal adaptations to 52

dry conditions, plant adaptations to 50

E

ecological relationships 82

ecosystem overfishing 249

ecstasy 37

effector 20

effluent 236

egested substances 62

electron microscopes 100

embryo 26, 156

embryo screening 164

embryo transplants 76

embryonic stem cells 156

energy and biomass 62–63

energy flow in food chains 244

energy flow through mammals and birds 246

energy stores 114

environment, adapting to 52–53

environment, effect of changing 58–59

environment, plants adapting to 50–51

environment manipulation for crops 118–119

environmental data collection 233

enzyme technology 142

enzymes 10, 100, 137, 138–139, 207

enzymes, bacterial 144

enzymes, digestive 140

enzymes, domestic uses 144–145

enzymes, effect of pH 139

enzymes, effect of temperature 138

enzymes, industrial uses 142–143

epidemics 15

ethanol biofuels 234–235

ethanol, testing for in urine 209

ethene (ethylene) 31

ethical questions 157

evolution, natural selection 84–85

evolution, theory of 84, 86–87, 166

evolution of life 82–83

evolutionary relationships 83

evolutionary tree 167

exchange systems in humans 188–189

exchange systems in plants 192–193

excretion 62, 206

exercise 4–5

exercise, body temperature 22

exercising 148–149

extinctions 168–169

extracellular enzymes 140

extreme microorganisms 56–57

extremophiles 56

F

factory farming 246

faeces 62, 206

family trees 162

farming, intensive animal 246–247

fatigue 150

fats 4

feedback control 214

feeding the world 250–251

fermentation 57

fertilisation 155

fertility drugs 25
fertility treatment 26–27
fertilised egg 74
fish stocks 248–249
fishing 248–249
fishing, overfishing 249
fishing, sustainable 249
fixing carbon dioxide 68
flagella 56
follicle stimulating hormone (FSH) 24, 25
food, high-protein from the sea 248
food, protein-rich from fungus 242–243
food absorption 189
food chains 244
food miles 251
food production in forest areas 231
food production, improving efficiency 244–245
food security 250
food webs 63, 249
forensic science 159
fossil evidence for evolution 83, 166–167
fossil fuels 68
fruit ripening 31
fungi, protein-rich foods 242–243
Fusarium fungus 242–243

G

gall bladder 141
gamete 72, 74
gametes 155
gas chromatography 159
gas exchange 188
gas exchange in humans 190–191
gas exchange through leaves 192
gateway drug 38
gel for growing bacteria 12, 18
gene mutations 85
gene transfer 79
genes 72–73, 104, 158–159
genetic code, modifying 78–79
genetic diagram 160
genetic engineering 78
genetic material 100
genetic variation 72
genetically modified (GM) crops 80–81
genetically modified (GM) organisms 78
germs 9
glands 20, 24

global warming 232–233
glucagon 216
glucose 22, 100, 206
glucose, controlling blood levels 216–217, 218
glucose, uses of 114–115
glycogen 148
Government directives for renewable energy 235
gravitropism 28–29
greenhouse effect 232
greenhouse gases 232
greenhouses 116
ground cover 125

H

Haber process 142
haemodialysis 210–211
haemoglobin 199
haemoglobin-based oxygen carriers (HBOCs) 199
hand-controlled ventilators 191
HBOCs (haemoglobin-based oxygen carriers) 199
heart 194
heart, artificial 201
heart rate 148
heart surgery 200–201
heart valves 194
heart valves, replacing 200–201
heat exhaustion 22–23
heat, animal adaptations to 52
heatstroke 22–23
herbicide 80, 229
heroin 37
homeostasis 22, 206
hormones 24, 216
host mothers 76
household waste 228
human activities and waste 228–229
humans, exchange systems 188–189
hydrothermal vents 56
hydrotropism 29
hygiene 8
hyphae 242

I

illness 12
immune system 213

immunisation (vaccination) 12–13
immunisation (vaccination) programmes 16–17
immunity 10
immunosuppressant drugs 213
impulses 20
in vitro fertilisation (IVF) 26
incubation 19
incubator 18
infections, controlling 14–15
ingesting pathogens 10
inheritance 160–161
inheritance, disorders 162–163
inherited characteristics 72
inoculating loop 19
inoculation into media 19
insoluble substances 114
inspiration 190
insulation, animals 52
insulation, plants 50
insulin 216
insulin, measuring the dose 219
insulin pumps 218
intensive animal farming 246–247
intercostal muscles 190
intra cytoplasmic sperm injection (ICSI) 27
in-vessel composting 67
ion absorption by plants 192
ionising radiation, sterilising with 18
ions 185
iron lung 190–191
IVF (in vitro fertilisation) 164

K

kidney failure 210–211
kidney transplants 212–213
kidneys 23, 208–209

L

lactic acid 150
lagoons, biogas production 240–241
land use for food 245
landfill sites 66
landfill sites, biogas production 240
leprosy 35
lichens 60
life on Earth 167
lifestyle 4

light intensity 112
limiting factors 112–113
lipase 140
lipids 115
liver 141

M

malnutrition 4
mass extinctions 168
maturation of human eggs 26
mean average 122
median average 122
medicine 12
meiosis 155
Mendel, Gregor 161
menstrual cycle 24
metabolic rate 4
metabolism 4–5
methanogens 66
microorganisms 8
microorganisms, extreme 56–57
mineral ions 206
mineral replacement 166
minerals 4
mitochondria 100
mitosis 154–155
MMR vaccine 16
MMR vaccine, controversy about 17
mode average 123
modelling climate change 232
monitoring body temperature 214
motor neurones 20
moulds 166
moulting of fur 52
mouth 141
MRSA 14
mummification 166
muscles, intercostal 190
mutations 14, 85
mycoproteins 243

N

natural selection 14, 84–85, 168
negatively gravitropic parts of plants 28
negative-pressure ventilators 190–191
nervous system 20–21
net movement of particles 107
neurones 20

new species 170–171
nitrates 115
nucleus of cells 20, 100
nutrients 64

O

obesity 217
oestrogen 24
organ donors 212–213
organ systems, specialised 108–109
organelles 100
organic substances 64
organic waste 66
organs 101
osmosis 184–185, 206
osmosis, reverse 185
ovaries 155
overfishing, ecosystem 249
ovum (human egg) 24
ovum (human egg), fertilised 74
ovum (human egg), maturation 26
oxygen debt 151
oxygen transportation by blood 199
oxygenation of blood 188
oxyhaemoglobin 199

P

palisade mesophyll cells 102
pancreas 141, 216
pandemics 15
partially permeable membranes 107
pathogenic microorganisms, killed during composting 67
pathogens 8–9, 213
pathogens, defences against 10–11
peat bogs 231
penicillin 12
permeable substances 29
pesticides 229
Petri dishes 18
PFCs (polyfluorocarbons) 199
pharmaceuticals 230
phloem 202–203
photosynthesis 28, 62, 102, 110–111
photosynthesis, enhancing 116–117
photosynthesis, glucose production 114–115
photosynthesis, rate of 112–113
phototropism 28

physical conditions 50
physiotherapy 164
placebo 34, 41
plankton 110
plant adaptations 50–51
plant cells 102–103
plant hormones (auxins) 28
plant organs 109
plant responses 28–29
plants, exchange systems 192–193
plants, transport systems 202–203
plasma (of the blood) 208
plasmolysis 185
platelets 198
poliomyelitis 190
pollination 81
pollution 60–61, 229
pollution indicators 60–61
polydactyly 162
polyfluorocarbons (PFCs) 199
polypeptides 136
polytunnels 116
population growth 228
positively gravitropic parts of plants 28
positively hydrotropic parts of plants 29
positively phototropic parts of plants 28
positive-pressure ventilators 190, 191
potometer 193
predation 168
pregnancy 24–25
producers 57
products 138
protease 140
protein-rich food from fungus 242–243
proteins 4, 100, 136–137
pulmonary artery 194
pulmonary vein 194
pulse 196
Punnett square 160
pure culture 19
pure-breeding lines 161
pyramid of biomass 62

Q

quadrats 124

R

radioactive dating 166
radioactive isotopes 203

rainforests, plant adaptations to 50
random sampling 125
randomised controlled trial 35
receptors 20
recessive alleles 160
recessive disorders 163
recreational drugs 36–37
recreational drugs, harm caused 38–39
recycling 66–67
recycling, natural 64–65
reduction in biodiversity 230
reflex actions 21
rehydration 186
relay neurones 20
renewable energy 235
reproduction, asexual 74–75, 155
reproduction, sexual 74–75
resistance to antibiotics 14
respiration 62, 206
respiration, aerobic 146–147
respiration, anaerobic 150–151
reverse osmosis 185
ribcage 190
ribosomes 100
root hair cells 102
rooting powders 31
rotting 64
rubbish disposal 66
runners (on plants) 74

S

sampling 124, 125
screening for disorders 164–165
sea, high-protein food 248
sea level rise 233
selective weedkillers 30
Semmelweiss, Ignaz 8
sensory neurones 20
sequestration 233
sex chromosomes 160
sexual reproduction 74–75
side effects of drugs 34
sieve plates 202, 203
simvastatin 35
slimming (dieting) 6–7
slimming drugs 6
small intestine 141
soil, plant adaptations to 51
soluble substances 106
solutes 202

specialise 76
specialised cells 100–101
specialised organ systems 108–109
speciation 170–171
species 82, 84
species, new 170–171
spinal cord 20
sports drinks 186, 187
stable natural communities 65
staple crops 250
staple diets 114
starch 110
starch, as energy store 114
statins 35
stem cells 156
stents 200
sterilisation 18–19
steroids 40–41
stomach 141
stomata (singular: stoma) 50, 192
storage organs 74
strains of microorganisms 18
substrates 138
succession 126–127
superbugs 14–15
surgery on the heart 200–201
survival of organisms 54–55
sweating 215
symptoms 10, 164
synapse 21
systematic sampling 125

T

Tamiflu 15
temperature (of human body) 22–23
temperature, controlling 214–215
territories 55
testes 40, 155
testosterone 40
test-tube babies 26
thalidomide 35
theory of evolution 84, 166
theory of evolution, development 86–87
thermoregulatory centre 214
thorax 190
tissue culture 76
tissue types 213
tissues 108
toxicity of drugs 34
toxins 10

trace fossils 166
transects 126
transgenic organisms 78
transpiration 192–193
transplantations of organs 212–213
transport systems in plants 202–203
transportation by blood 198
transportation of oxygen 199
trophic level 63
tropical deforestation 230–231
tubules 208
turgor 185

U

ultraviolet radiation, sterilising with 18
unicellular organisms 104
unidirectional light 28
urea 206
urea, production of 208
urine 23, 206
urine, production of 208
urine, testing for ethanol content 209
uterus 24

V

vaccination (immunisation) 12–13
vaccination (immunisation)
   programmes 16–17
vaccines 12
vaccines, concerns about 16–17
vacuole 102
valves of the heart 194
valves of the heart, replacing 200–201
variation, categoric 72
variation, continuous 72
variation, genetic 72
variegated leaves 110
vascular bundles 202–203
veins 194, 196
vena cava 194
ventricle 194
ventilation 190
ventilators, hand-controlled 191
ventilators, negative-pressure 190–191
ventilators, positive-pressure 190, 191
villi 188, 189
viruses 9
viruses, changing 15
vitamins 4

W

waste, production of biogas 238–239
waste disposal 66
waste from human activities 228–229
water, transport through plants 202–203
water absorption by plants 192
water budget of human body 23
water security 250
weed plants 80
weedkillers 30
white blood cells 10–11, 198
wildlife 229
windrow composting 67
world food supply 250–251

X

xylem vessels 103

Y

yeast 105
yields of food crops 59, 80, 116

Z

zero-grazing dairy systems 247
zygotes 108